CHROMATOGRAPHIC METHODS

Chromatographic Methods

R. Stock
Head of Department of Physical Sciences
Trent Polytechnic, Nottingham

C. B. F. Rice
Principal Lecturer in Chemistry
Liverpool Polytechnic

CHAPMAN AND HALL

and SCIENCE PAPERBACKS

First published 1963
Second edition 1967
Third edition 1974
Reprinted 1977
Chapman and Hall Ltd
11 New Fetter Lane, London EC4P 4EE

© 1963, 1967, 1974 R. Stock and C. B. F. Rice

Typeset by Santype Ltd (Coldtype Division)
Salisbury, Wiltshire
and printed in Great Britain by
Fletcher & Son Ltd, Norwich

ISBN 0 412 10560 8 (cased edition)
ISBN 0 412 20810 5 (Science Paperback edition)

Distributed in the U.S.A.
by Halsted Press, a Division
of John Wiley & Sons, Inc., New York

Contents

Preface to First Edition

The various methods of separating mixtures which are grouped under the general name 'chromatography' are now well known and widely used. Since the inception of chromatography as a column technique in 1903, the principal landmarks in its progress have been its virtual rediscovery in 1930, the invention of synthetic ion-exchange resins in 1935, the introduction of paper chromatography in the early 1940's, and finally the development of gas-solid and gas-liquid chromatography in the late 1940's and early 1950's. Subsequent expansion in the use of chromatographic methods has been rapid and continuous, with the result that in the last fifteen years a substantial volume of literature on the subject has appeared, dealing not only with particular separations but also in much specific detail with improvements in technique.

Many specialist books have been published. Some are concerned only with particular aspects of the subject. Others are essentially literature surveys which are usually very comprehensive (though somewhat uncritical), and hence rather formidable to someone seeking an introduction to chromatography.

The present book aims to present a short account of techniques in current use. Emphasis has been given to paper and gas chromatography because they are now the most widely used methods, although thin-layer chromatography has recently become extremely popular. The treatment is dictated by the need to stress practical considerations, and is designed to show how the various methods complement one another. Details are given of some model experiments based on the experience of the authors in teaching the subject. The inclusion of student experiments in gas chromatography is regarded as an important development. Zone electrophoresis, though not strictly a chromatographic technique, is nearly enough related to be included. An attempt has been made to give some guidance as to the choice of method for a particular purpose, and to assess the relative values of the different procedures.

The authors gratefully acknowledge permission to reproduce published information, diagrams, or photographs from: B.D.H. Ltd., The Dow Corporation, Evans Electroselenium Ltd., Mr. G. F. Harrison of Associated Octel Co. Ltd., Dr. C. S. Knight of W. R. Balston & Co. Ltd., Dr. S. W. S. McCreadie and Dr. A. F. Williams of Nobel Division, I.C.I. Ltd., The Permutit Co. Ltd., H. Reeve Angel & Co. Ltd., the Journal of Applied Chemistry, the Journal of Chemical Education, and Nature. Thanks are also due to Dr. S. J. Gregg for much helpful criticism and discussion, to Dr. G. Skirrow for suggesting certain of the gas chromatography experiments, and, finally, to Mrs. R. Wiggins for preparation of the typescript.

College of Technology   R. S T O C K
Liverpool   C. B. F. R I C E

PREFACE TO THIRD EDITION

An evaluation of the literature relating to chromatography by Janak (Janak, J., *J. Chromatog.,* 1973, 78, 117) has shown that since 1967 the number of research papers on paper and thin-layer chromatography has fallen, the number on gas chromatography, after an initial fall, has started to rise again, and the number on liquid chromatography has shown a spectacular increase. The number of research papers published in a particular period is not necessarily an exact measure of the extent to which a method is actually used during the time in question, but it is probably indicative of trends.

The increase in interest in liquid chromatography on columns, although due in part to the continued development of gel chromatography, is due mainly to the application of the methods of gas chromatography in what is now called high efficiency liquid chromatography, accompanied by increased sophistication in instrumentation and improvements in manufacture of chromatographic media.

These trends are reflected in the third edition of this book. The sections on gas chromatography and gel chromatography have been rewritten and expanded, and a section on high efficiency liquid chromatography has been included, but the remaining section, apart from minor revisions and corrections, are substantially as in the second edition. Throughout the book SI units are now used and references to commercial products have been brought up to date. The bibliography has been made more selective and the list of visual aids for teaching has been extended.

February 1974

R. STOCK
Trent Polytechnic, Nottingham
C. B. F. RICE
Liverpool Polytechnic

Chapter One

Introduction

Classification of chromatographic methods

Chromatography is the name given to a particular family of separation techniques of great effectiveness. The original method was described in 1903 by Tswett, who used it for the separation of coloured substances, and the name chromatography stems from this. However, the limitation to coloured compounds never really obtained, and most chromatographic separations are nowadays performed on mixtures of colourless substances, including gases.

Like fractional distillation, chromatography relies on the relative movement of two phases, but in chromatography one is fixed and is known as the *stationary* phase; the other is known as the *mobile* phase. Chromatographic methods may be classified first according to the nature of the mobile phase and, second, according to the nature of the stationary phase. The mobile phase may be a liquid or a gas, and the stationary phase may be a solid or a liquid. There are thus four main sub-divisions of the chromatographic process, as set out in Table 1.1. The system is called *adsorption* chromatography if the stationary phase is a solid, and *partition* chromatography if it is a liquid.

There are several variants of the four main types, as shown below:

1. *Liquid-solid*
   'Classical' adsorption chromatography
   Thin-layer chromatography
   Ion-exchange chromatography
   Gel chromatography

Table 1.1

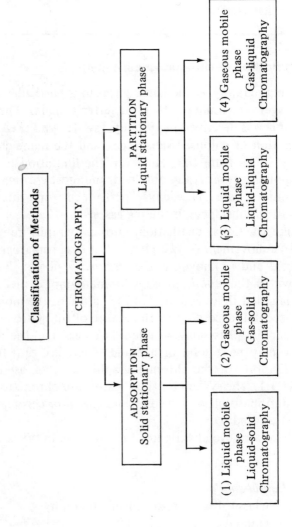

| Classification of Methods |

CHROMATOGRAPHY

ADSORPTION
Solid stationary phase

PARTITION
Liquid stationary phase

(1) Liquid mobile phase
Liquid-solid Chromatography

(2) Gaseous mobile phase
Gas-solid Chromatography

(3) Liquid mobile phase
Liquid-liquid Chromatography

(4) Gaseous mobile phase
Gas-liquid Chromatography

2. *Gas-solid*
    Gas-solid chromatography

3. *Liquid-liquid*
    'Classical' partition chromatography
    Paper chromatography

4. *Gas-liquid*
    Gas-liquid chromatography
    Capillary-column chromatography

To illustrate the technique two typical chromatographic experiments will be outlined; the first involving adsorption and the second partition. From these accounts the essential unity of the methods will emerge.

Adsorption chromatography

To separate a mixture, a column such as that illustrated in Fig. 1.1 (*a*) may be used. It is packed with an active solid such as alumina – the stationary phase – covered with a solvent such as hexane – the mobile phase. A small sample of the mixture applied to the top of the column forms a band of adsorbed material – Fig. 1.1 (*b*). When solvent is allowed to flow through the column it carries with it the components of the mixture. The rate of movement of a given component depends on how much it is retarded by adsorption on the column packing. Thus a weakly adsorbed substance travels more rapidly than a strongly adsorbed one. It can be seen that if the differences in adsorption are sufficiently great there will be complete separation. This is the stage shown in Fig. 1.1 (*d*) which is known as complete development.

The above is an example of liquid-solid chromatography. Separation is caused by differences in the adsorption forces between the various components of the mixture and the stationary phase. These forces may be either van der Waals'

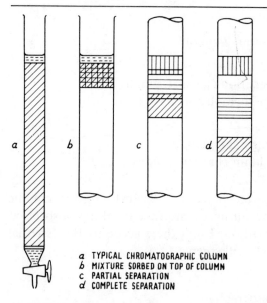

a TYPICAL CHROMATOGRAPHIC COLUMN
b MIXTURE SORBED ON TOP OF COLUMN
c PARTIAL SEPARATION
d COMPLETE SEPARATION

Fig. 1.1. Separation of a mixture by column chromatography

forces, as in the case of alumina just described, or electrostatic forces, as in the separation of ions by ion-exchange chromatography, where the stationary phase is an ion-exchange material, or a column packing of a suitable porous nature may be used to separate solutes according to their molecular size – as in gel chromatography. Thin-layer chromatography is a special example of adsorption chromatography where the adsorbent is spread out as a thin film on a glass plate, plastic film or metal foil instead of being packed in a cylindrical column. The nature of active solids and ion-exchange materials is discussed in Chapter 2.

Partition chromatography

In separations by partition chromatography on a column the technique is very similar to that just described. The principal difference is the nature of the column packing, which is now a porous material, such as kieselguhr, coated with a layer of liquid — frequently water. The stationary phase is the liquid layer and the solid merely serves as a support for it. If the mobile and the stationary phases are both liquids an example of liquid-liquid chromatography is obtained.

The rate of movement of a given component of the mixture depends no longer on adsorption, but on its solubility in the stationary phase; more soluble substances travel more slowly down the column than the less soluble. During their passage the substances undergo *partition* between the two phases, and separation occurs because of the differences in the partition coefficients.

Paper chromatography is a special case of partition, in which paper sheets take the place of the packed column.

All separations by chromatography depend on the fact that the substances to be separated distribute themselves between the mobile and the stationary phases in proportions which vary from one substance to another. The manner in which the substances are distributed is most conveniently discussed by referring to the 'sorption' isotherm.

Sorption isotherms

The amount of a particular substance taken up — 'sorbed' — by the stationary phase depends on the concentration in the mobile phase. The curve obtained by plotting the amount sorbed against concentration, at constant temperature, is the 'sorption isotherm'. The shape of the isotherm is one of the most important factors governing chromatographic behaviour.

The term 'sorption' includes the processes of *ad*sorption and *ab*sorption. In the present work adsorption refers to the increase in concentration (in excess of that in the mobile phase) at the interface between the mobile and the (solid) stationary phases. Absorption refers to the dissolution of a substance from the mobile phase into the liquid stationary phase, that is, partition.

Figure 1.2 (*a*) shows a typical adsorption isotherm, and Fig. 1.2 (*d*) and (*g*) the characteristics of the chromatographic band with which it is associated. In order to obtain the chromatographic band it is assumed that a column is packed with the stationary phase and a sample of the substance is then carried through it by the flow of mobile phase as in the experiments described above. The bands in Fig. 1.2 are shown at some stage in their passage down the column. Bands with sharp fronts and long diffuse tails are said to

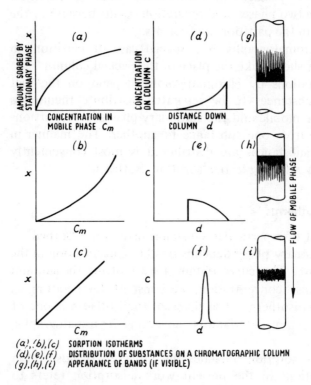

(*a*),(*b*),(*c*)   SORPTION ISOTHERMS
(*d*),(*e*),(*f*)   DISTRIBUTION OF SUBSTANCES ON A CHROMATOGRAPHIC COLUMN
(*g*),(*h*),(*i*)   APPEARANCE OF BANDS (IF VISIBLE)

Fig. 1.2. Sorption isotherms and chromatographic behaviour

exhibit 'tailing', and the steeper the isotherm near the origin, the longer and more diffuse the tail. Tailing is the principal disadvantage of adsorption chromatography.

The isotherm in Fig. 1.2 (*b*) is associated with 'fronting' – Fig. 1.2 (*e*) and (*h*). This type of isotherm is rare in adsorption systems, but may occur more frequently in partition work.

When the isotherm is linear, as in Fig. 1.2 (*c*), chromatographic bands which are compact and symmetrical are obtained – Fig. 1.2 (*f*) and (*i*). Such behaviour is characteristic of partition systems, but may also occur in adsorption.

Non-equilibrium effects may also give rise to fronting or tailing, but they may be minimised by paying careful attention to the operating conditions.

An essential feature of chromatography is that the mixture of substances initially forms a band on the column, either adsorbed or absorbed by the stationary phase. To achieve separation of the components the band has to be subjected to further treatment, which may be one of three kinds: *elution, frontal analysis* or *displacement development analysis*. All these techniques are due, at least in part, to Tiselius and his co-workers.

Separation techniques
Elution analysis

The nature of the elution technique has been indicated in the two examples given earlier. It is easily the most important method, and is indeed the only one used extensively in partition chromatography. In elution the flow of the mobile phase (the 'eluting agent' or eluent) is continued until the mixture is completely separated into its components. It is important that the mobile phase chosen shall be unaffected by the stationary phase, or at most only interact weakly with it, otherwise displacement development may occur (see below). Elution was first used by Tswett in 1903 for the separation of leaf pigments, which, because of their colour, were readily located on the column. Colourless substances may be located by, for example, their fluorescence in ultra-violet light. Once separation has occurred the column packing may be extruded and cut up with a knife, and the portions containing the various components extracted with solvents. Alternatively, and more usually, the flow of eluent may be continued until each component is flushed completely from the column; it may then be detected in the effluent or *eluate* by suitable chemical or physical tests. The amount of liquid needed to elute a given substance is known as the *retention volume* of that substance, and the time taken as the retention time.

In 'stepwise' elution several eluting agents are used in succession, arranged such that each is more effective than the one preceding; that is, able to elute a given component from the column more readily — see p. 23.

Another useful modification, known as gradient elution analysis, is particularly valuable in adsorption chromatography (including ion-exchange) because it reduces tailing; it is also used in liquid-liquid chromatography when slow-moving bands are encountered. An eluting agent stronger than the one currently in use is added to the column in such a way (p. 24) that a continuous concentration gradient is established down the column; the rear part of each chromatographic band is then always in contact with a more strongly eluting solution than is the front, and each band is in a more powerful eluting medium than the band preceding. The effect is thus twofold: each band is compressed into a narrower region of the column and the distance between bands is decreased. The method has much in common with the displacement development technique which will shortly be described (Hagdahl, Williams and Tiselius[1] consider elution, gradient elution and displacement development to be different aspects of the same basic technique).

Tailing, which frequently occurs in adsorption chromatography when the elution technique is used, is undesirable because it increases the width of the bands and, hence, their tendency to overlap. The sharp fronts associated with tailing are, however, a desirable feature. In displacement development and frontal analysis advantage is taken of the sharp fronts, but tailing is eliminated, so that each component is confined to a narrow section of the column.

Frontal analysis

In frontal analysis a solution of the mixture is added *continuously* until the column is saturated. If the concentration of the solution emerging from the bottom of the column is measured continuously a stepped curve is obtained (Fig. 1.3). Sharp steps will only be obtained if the system gives isotherms of the type in Fig. 1.2 (*a*), since only if the isotherms are of that shape will the chromatographic bands have sharp fronts. The method is therefore particularly suitable in adsorption chromatography.

In the example in Fig. 1.3 it is assumed that component *A* is the least, and component *C* the most strongly adsorbed by

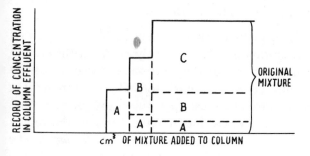

Fig. 1.3. Frontal analysis

the column packing. When the mixture is added continuously to the column, A will tend to run ahead of the other components and the first step will consist of pure A. Component B is more strongly adsorbed than A but less than C; thus it will tend to move away from the main body of the mixture but will not overtake A. This second step will thus consist mainly of B but will be contaminated with A because of the continued supply from the main body of the mixture. The third step will consist of the original mixture $A + B + C$, at first slightly depleted in A and B. The number of steps is thus equal to the number of components in the original mixture, but only the first step is a pure component. A complete separation cannot be achieved and, in this respect frontal analysis differs from all the other methods.

In principle it is possible to calculate from adsorption data the proportions of each component in the original mixture but in order to do so much preliminary calibration is necessary and quantitative measurements are only practicable with very simple mixtures. A typical application would be the estimation of a trace impurity in a nearly pure substance. The impurity would tend to be concentrated in front of the main constituent, provided of course that it was the less strongly adsorbed.

Displacement development analysis

Displacement development analysis may perhaps be regarded as a hybrid of elution and frontal analysis. As in the elution method, a small amount of the mixture is placed on top of the column, but instead of pure solvent, a solution of a substance more strongly adsorbed than any of the components of the mixture is added continuously to the top of the column. The substance is known as the *displacer*. The mixture then moves down the column at the same rate as the displacer is added and the mixture resolves itself into bands of pure components; the order of the bands is the order of the strength of adsorption on the packing material. Each pure band acts as a displacer for the component ahead of it, and the last and most strongly adsorbed band is forced along by the displacer.

The record of the concentration of material in the column effluent is shown in Fig. 1.4 (*a*), and it is seen to be similar to the record obtained in the frontal analysis method. The important difference is that the steps represent pure components. Such a record is better suited to quantitative work than the record obtained in frontal analysis.

Once again, for sharp steps the system should give isotherms of the type shown in Fig. 1.2 (*a*). It can be shown that the concentration of the displacer and the respective isotherms determine the concentrations of the various components (*A*, *B* and *C*) in the steps, in the manner shown in Fig. 1.4 (*b*). Curve *D* is the isotherm for the displacer. If a

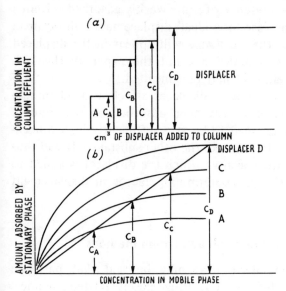

Fig. 1.4. Displacement analysis

concentration C_d is used, and the corresponding point on the isotherm is jointed to the origin, the line ('displacer' line) cuts the isotherms of C, B and A at points corresponding to the concentrations C_c, C_b and C_a which are then the concentrations in the displaced steps. If a component is so weakly adsorbed that its isotherm is not cut by the displacer line, under the conditions of displacement development it will travel ahead of the rest of the mixture and form an ordinary elution peak (Fig. 1.2 (*d*)).

Strictly speaking, the above conclusions are valid only when the isotherms obey the Langmuir equation ('Langmuir' isotherms). Also, for a given component to be displaced completely, it is necessary that its isotherm, when measured *in the presence of the displacer*, should be depressed so much from its usual position that the displacer line no longer cuts it. (In Fig. 1.4 (*b*) the isotherms of the displacer and the different components are assumed to have been measured separately in the presence of some weakly adsorbed solvent.) If this is not the case 'non-ideal' displacement will occur in which not all of the substance will appear in the displaced step. For a more detailed account the paper by Hagdahl, Williams and Tiselius[1] may be consulted.

Figure 1.5 shows how a displacer affects a band of substance which has previously been adsorbed on the column. The displacer is assumed to be at a sufficiently high concentration to be able to overtake the other substance. It will thus be evident that the speed at which the displacer is added to the column is the rate-determining factor in displacement analysis.

Comparison of partition and adsorption methods

As a rough generalisation it may be said that partition methods are better for separating closely related chemical types, such as the members of homologous series, and that

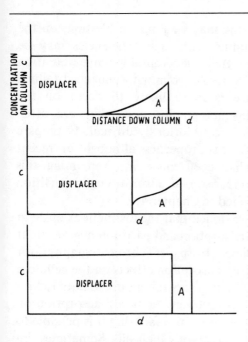

*Fig. 1.5. Displacer overtaking previously
adsorbed band*

adsorption methods are better for separating different chemical types because of the greater specificity of adsorption effects. The main advantage of partition lies in the relatively narrow chromatographic bands obtained because of absence of tailing, giving rise to more efficient use of columns – although, as already indicated, tailing can be reduced in adsorption, preferably by stepwise elution or gradient elution; displacement development and frontal analysis are recommended only for very specialised applications. Other methods for reducing tailing in adsorption chromatography, including 'straightening' of the isotherm are discussed on p. 243.

The capacity of an adsorption column, per unit volume, is frequently greater than that of a partition one, sometimes by a very large amount. This may be a mixed blessing: on the one hand shorter columns with high efficiencies may be obtained, but on the other, unacceptably large retention volumes (see p. 29) may result along with enhanced tailing. Again, chemical change is more likely to occur on the adsorbents used in chromatography since they are often efficient catalysts: also it is notoriously difficult to prepare adsorbents with consistent properties although in recent years much progress has been made in overcoming this problem. A further comparison of adsorption and partition chromatography is included in Chapter 4.

Finally, it may be mentioned that the distinctions made in Table 1.1 are not rigid; adsorption and partition may occur at one and the same time. In paper chromatography, for example, there are usually adsorption effects on the cellulose fibre. Again, one often encounters tailing in gas-liquid chromatography. The operation of van der Waals' adsorption on ion-exchange resins is shown by the fact that it is possible to separate non-ionic substances with their aid. Sometimes, indeed, it is not possible to say whether adsorption or partition is the more important mechanism (see Chapter 2).

Efficiency of chromatographic methods

It possible to achieve by chromatographic methods separations which are very difficult by other means. The reason is that during their passage through a chromatographic system small differences in sorptive behaviour of substances are multiplied many times. The greater this 'multiplication' factor the greater the separating power, and the less need be the differences in order to achieve separation.

Chromatography has much in common with fractional distillation, and the 'plate' theory of partition chromatography emphasises this relationship. A distillation column may be considered to consist of a stack of units, known as theoretical plates, in each of which equilibrium is established between the ascending vapour and the descending liquid. The larger the number of theoretical plates the more efficient the column. In a similar way, a chromatographic column may be considered as a collection of plates, a plate being defined as a section of column in which equilibrium is established between the mobile and the stationary phases.

The two types of plates are not strictly comparable, but roughly it may be said that one distillation plate is equivalent to between 10 and 30 chromatographic plates. Thus, to achieve a separating power equivalent to 100 distillation plates, a chromatographic column must possess up to 3000 plates. It is the advantage of chromatography that this can be achieved relatively easily, and, whereas the distillation column may take several days to reach its maximum efficiency, a chromatographic separation may be performed in a matter of minutes or hours. In gas-liquid chromatography it is possible to set up a column with a plate number in excess of 30,000; in fact capillary columns (see Chapter 4) have been described which possess 500,000 or more theoretical plates.

The importance of the plate number is that it gives a measure of the separating power of the column. In adsorption chromatography the plate theory is only applicable when the isotherm is linear; when curved isotherms are obtained it is not possible to express the separating efficiency in a simple manner. A more detailed account of the plate theory appears in Chapter 4 since this theory finds particularly wide application in gas chromatography.

Other advantages of chromatography

A notable advantage of chromatographic methods is that they are comparatively 'gentle', in the sense that decomposition of the substances being separated is less likely than in some other methods. This is an important consideration in the case of labile substances, often of biological origin. As mentioned above, however, catalysed decomposition may sometimes occur on an adsorption column. A further advantage is that only a very small quantity of the mixture is required so that analytical techniques involving chromatographic separations can be carried out on a micro or semi-micro scale. Indeed it is not possible to handle very large quantities by chromatography and this constitutes a disadvantage for preparative purposes. Chromatographic techniques are simple and rapid, and the apparatus required is cheap; complex mixtures can be handled with comparative ease.

The choice of method

Except when the technique to be used is obvious (such as gases by gas chromatography) the choice of method must be largely empirical because there is as yet no way of predicting the best procedure for a given separation, except in a few simple cases. It is usual to try the simpler techniques such as paper and thin-layer chromatography first because they can often provide a useful guide to the type of system that will ultimately prove successful if they themselves do not provide the answer directly. The more sophisticated techniques can then be applied as necessary; the list that follows is a rough guide.

1. Substances of similar chemical type — Partition chromatography

2. Substances of different chemical type — Adsorption chromatography

3. Gases and volatile substances — Gas chromatography

4. Ionic and inorganic substances — Ion-exchange chromatography on columns, paper or thin layers; zone electrophoresis (see p. 170)

| 5. Ionic from non-ionic substances | Ion-exchange or gel chromatography |
| 6. Biological materials and compounds of high relative molecular mass | Gel chromatography; electrophoresis |

In the event of difficult separations, when the simpler methods prove inadequate, high efficiency liquid chromatography (HELC) may provide the answer. This technique is described in Chapter 2 (p. 85).

REFERENCE

1. Hagdahl, L., Williams, R. J. P., Tiselius, A., *Arkiv Kemi*, 1952, **4**, 193.

Chapter Two

Liquid-phase Chromatography on Columns

In this chapter we will consider the principal methods of chromatography in which a packed column is used with a liquid moving phase. We are here grouping together a number of different physical systems, and the common feature is the practical technique which is used; the broad sub-divisions were described in Chapter 1. The solid column packing may be itself the stationary phase (adsorption or ion-exchange chromatography) or the packing may be the support for a liquid stationary phase (partition chromatography). In practice the term 'adsorption chromatography' is restricted to cases where adsorption is principally by van der Waals forces, and 'ion-exchange chromatography' is used when the packing is an ion-exchange material and adsorption is principally by electrostatic forces. The newer technique of gel chromatography (gel filtration), where the stationary phase is a porous gel and the separation is according to molecular size, is included in this chapter because the methods used are commonly those of conventional column chromatography.

Although chromatography on columns was the original method, the more convenient techniques of partition chromatography on paper and adsorption chromatography on thin-layer plates have until recently been more widely used. Automation of ion-exchange column chromatography, the introduction of gel chromatography, and, above all, the use of the methods of gas chromatography in high performance liquid chromatography, have brought about a resurgence of interest in column methods, and liquid chromatography on columns may become again the most widely used technique.

Adsorption columns

A simple adsorption column was described in Chapter 1. It was the type of column first used for chromatographic separations of the kind familiar today. After its introduction (in about 1903 by Tswett) there was little or no application of adsorption chromatography until it was again successfully employed in 1931 by Kuhn, Winterstein and Lederer[1] for the separation of xanthophyll pigments on columns packed with calcium carbonate. After that time much more use was made of adsorption chromatography, and there were minor improvements in technique. It was left to Tiselius and his school, however, to make most of the major advances, such as the perfection of the elution method of separation and the devising of apparatus for the continuous analysis of the eluate by, for example, measuring changes in the refractive index. One advantage is that this makes convenient the use of such adsorbents as active carbon as column packings because it is no longer necessary to detect the separated components while they are still on the column. Other advances, such as frontal analysis, displacement development, and gradient elution analysis, also invented by Tiselius and his co-workers, aim at the reduction or elimination of the 'tailing' always associated with adsorption chromatography, and help to achieve more efficient columns. A brief description of these methods has already been outlined in Chapter 1, and further experiments employing them will be described below and in Chapter 6. Gradient elution analysis seems to be the most generally useful of these modifications, and it is also employed extensively in partition and ion-exchange chromatography.

Columns

Chromatographic columns can be obtained commercially with such refinements as spring-loaded stopcocks and sintered-glass discs. They can also be constructed quite simply (and cheaply) from glass tubing. Sometimes an old burette will do. The size of the column depends on the quantity of material to be separated. The smallest columns are only a few millimetres in diameter and a few centimetres long, while the largest may be several centimetres in diameter and of correspondingly greater length. Some model experiments are described in Chapter 6, where column sizes are specified. The lower end of the column is drawn out so that it can be connected to a stopcock or a piece of flexible tubing which can be pinched with a screw-clip. The adsorbent can be supported on a plug of glass wool or on a porous plate which rests on lugs. The latter is preferable if it is desired to extrude the column after separation (Fig. 2.1).

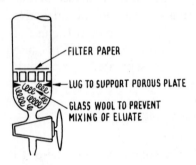

FILTER PAPER

LUG TO SUPPORT POROUS PLATE

GLASS WOOL TO PREVENT MIXING OF ELUATE

Fig. 2.1. Supporting the column packing

Packing the column

It is not easy to obtain a uniformly packed column, but it is essential to try in order to achieve maximum efficiency. Practice may be necessary before completely satisfactory results are obtained. Irregular packing is due largely to the separation of particles of different sizes and is the cause of distorted chromatographic boundaries, which in turn make the bands take up more room than they should on the column. Channelling is usually caused by the inclusion of air bubbles during packing. To prevent these effects so far as possible the packing material should be slurried with the solvent and poured in a thin stream into the tube, already about one-third full of solvent. If the adsorbent is allowed to settle gradually – which can be arranged by maintaining gentle agitation while there is a solvent flow through the column – reasonably homogeneous packing will follow. If the particle size of the adsorbent is uniform, it is easier to get homogeneous packing. On no account should any part of the column be allowed to run dry, during packing or during a separation.

Application of the sample

It is important to apply the sample to the top of the column as evenly as possible, in as concentrated a solution as possible (while avoiding precipitation), and disturbance of the column packing should be avoided because that will lead to distorted bands. Some workers protect the top of the column with a piece of filter paper or a thin layer of clean sand. Application of the sample can be made with a small pipette (Fig. 2.2). The tip of the pipette is placed against the column wall (a

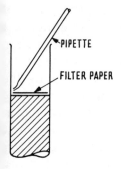

Fig. 2.2. Sample application

slight bend in the tip will make this easier) just above the surface of the adsorbent (just exposed); while the liquid is slowly draining from the pipette the tip is run round the inside of the column, care being taken not to touch the packing; any sample remaining on the walls can be carefully washed on to the column in a similar way, using pure solvent. When all the sample has been adsorbed on the top of the column the vacant space above it can be filled with solvent and the column allowed to run; the supply of solvent can be replenished from a separating funnel. Alternatively a non-draining device may be used (Fig. 2.7 (*b*)). Other methods of applying the sample are mentioned in the section on partition chromatography (p. 40).

Simple elution

By simple elution is meant the operation of the chromatographic column by allowing a solvent mixture of unvarying composition to run through the column until separation is complete. This solvent need not be the same as that used to make up the column initially, but if it is too strongly adsorbed by the column packing displacement development may occur. If the separated constituents of the mixture can be observed on the column (either by their colour, their reaction with an indicator previously or subsequently applied to the column, or, perhaps, their fluorescence in ultra-violet light), the run can be stopped. The contents of the column can now be extruded and the separated constituents extracted by means of suitable solvents. Extrusion of the column intact is not easy, and is rendered unnecessary by the use of a transparent nylon tube as the column container. When the separated constituents have been located the whole tube is cut into sections and the separate parts of the column are removed. An alternative and more commonly used method is to allow the column to run until the separated components can be detected in the column effluent (eluate). The latter method must be used if instrumental techniques are to be used to monitor the separation (p. 84). Generally speaking the more slowly the column is run, the sharper the chromatographic bands. That is because the system will be more nearly at equilibrium and there will be less band spreading due to finite rate of adsorption. Again, the more finely divided the adsorbent the more rapidly will equilibrium be attained, although the fine particles will tend to impede the solvent flow. To accelerate the solvent flow when it becomes too slow it is better to apply pressure to the top of the column by means of compressed gas rather than to reduce pressure at the bottom. Reduction of pressure is liable to cause channelling by allowing dissolved gases to come out of solution.

Stepwise or fractional elution

If only one solvent is used ready elution of only some of the components of the original mixture from the column may result. To remove those which are more firmly held a stronger eluting agent will be required. Sometimes it may be necessary to use several different solvents of gradually increasing strength for the successive desorption of different components. This is known as stepwise elution. It has the advantage that sharper separations may be obtained than if only one strongly eluting solvent, capable of moving even the most firmly bound of the components of the mixture, is used — apart from the possibility of displacement development. One danger of this technique, however, is that a given compound may give rise to more than one peak by appearing in the eluates of successive steps.

It is possible to construct so-called 'eluotropic' (or

'elutotropic') series in which solvents are arranged in order of their eluting power. Some of these are considered below under the heading 'Solvents'.

Gradient elution analysis

The technique of gradient elution analysis was first described in detail by Alm, Williams, and Tiselius[2]. It involves the use of a continuously changing eluting medium. For example, if water is the solvent initially in use on the column and ethanol is a stronger eluting agent, application of the gradient elution method involves the addition to the column of a solution of ethanol-in-water of gradually increasing strength. The effect of this gradient is to elute successively the more strongly adsorbed substances and at the same time to reduce tailing. This means that the chromatographic bands will tend to be more concentrated and thus occupy less of the column. This desirable effect may be ascribed to the 'straightening' of the isotherms by the concentration gradient; that is, the adsorption isotherms are becoming more nearly linear. It is not difficult to understand why this should happen; the use of a concentration gradient ensures that the tail of a particular chromatographic band is always in contact with a more concentrated solution (therefore more strongly eluting) than the front. The tail will therefore tend to move more rapidly and to catch up the front. In Fig. 2.3 is shown a simple apparatus for obtaining the gradient; the whole device may be under pressure if necessary. The flask B, which contains a magnetic stirrer, initially contains the pure solvent used to make up the column, and the solvent whose concentration is to be increased is added from A at rate R_1. R_2 is the rate at which the mixture is added to the column.

The rates R_1 and R_2 can be adjusted to give concentration curves similar to those depicted in Fig. 2.4. Type I curves (R_2

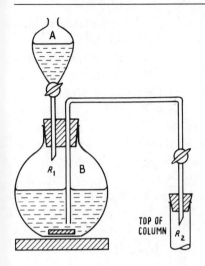

Fig. 2.3. Mixing device for gradient elution

greater than $2R_1$) give the best separations with the minimum tailing. By using the equation

$$C = 1 - \left(\frac{a}{a + bt} \right)^{1/b}$$

in which $a = V_0/R_1$, where V_0 = starting volume of pure solvent in vessel B, and $b = 1 - R_2/R_1$, the concentration C

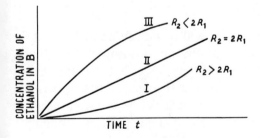

Fig. 2.4. Concentration gradients

of the stronger eluting agent in B can be calculated at any time t. If one uses an arrangement, as shown in Fig. 2.5, in which the two vessels are identical in size and shape one has the situation where $R_2 = 2R_1$ and thus obtains a linear concentration gradient. By altering the relative sizes of the vessels one can vary the ratio of R_1 to R_2. It is unwise to use too steep a gradient because this may lead to the production of two elution peaks from one substance. Shallow concentration curves can conveniently be obtained by adding to the mixing chamber a solution instead of pure solvent. An example would be the use of aqueous ethanol in vessel A in Fig. 2.3 and water in vessel B.

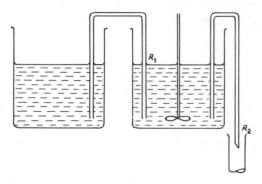

Fig. 2.5. Production of a linear gradient

Solvents

The initial choice of solvent will obviously be influenced by the nature and solubility of the mixture, but, other things being equal, it is often better to choose a solvent of indifferent eluting power so that stronger eluting agents can be tried successively. By 'strength' of the eluting agent is meant the adsorbability on the column packing. Generally, for polar adsorbents such as alumina and silica gel, the strength of adsorption increases with the polarity of the adsorbate. For carbon the order is reversed. One of the first 'eluotropic' series was recorded by Trappe[3] who found that the eluting power of a series of solvents for substances adsorbed in columns such as silica gel *decreased* in the order:

pure water > methanol > ethanol > propanol > acetone > ethyl acetate > diethyl ether > chloroform > dichloromethane > benzene > toluene > trichloroethylene > carbon tetrachloride > cyclohexane > hexane.

This order is also the order of decreasing dielectric constant. The eluting power of different solvents has also been studied by Williams[4] *et al.*, who found that on an active carbon column the eluting power for amino-acids and saccharides decreased in the order:

ethyl acetate > diethyl ether > propanol > acetone > ethanol > methanol > pure water.

This order is of increasing polarity or decreasing chain length of homologues. The reverse order was true for alumina and silica gel. Other series have been given by Strain[5], Bickoff[6], and Knight and Groennings[7]. With the aid of these series it is possible to select a solvent with the appropriate eluting power.

The purity of the solvents obtained should be as high as possible and, if necessary, further purification should be undertaken. Sometimes a useful purification can be achieved by running the solvent through a column of the adsorbent to be used.

Adsorbents

Many solids have been used as adsorbents (Tswett himself tried over 100 different compounds) some of which are listed in Table 2.1 with the sorts of compounds separated with their aid.

It may be noted that in the table there is little reference to inorganic separations, which are usually more conveniently carried out with the aid of ion-exchange resins or by paper chromatography.

Table 2.1   *Adsorbents for chromatography*

Solid	Used to separate
Alumina	Sterols, dyestuffs, vitamins, esters, alkaloids, inorganic compounds
Silica gel	Sterols, amino-acids
Carbon	Peptides, carbohydrates, amino-acids
Magnesia	Similar to alumina
Magnesium carbonate	Porphyrins
Magnesium silicate	Sterols, esters, glycerides, alkaloids
Calcium hydroxide	Carotenoids
Calcium carbonate	Carotenoids, xanthophylls
Calcium phosphate	Enzymes, proteins, polynucleotides
Aluminium silicate	Sterols
Starch	Enzymes
Sugar	Chlorophyll, xanthophyll

Table 2.1 is not meant to be exhaustive. Probably the most popular and generally useful adsorbent has been alumina. Many other solids have been used including those more usually associated with a different mechanism of separation, for example, ion-exchange resins, which have adsorptive capacities for non-ionic materials; the substances oestriol, oestradiol, and oestrone have been separated on the ion-exchange resin Amberlite IRC-50[8], a process which does not involve ion-exchange. More recently, cross-linked polystyrene, the basis of most ion-exchange resins, has been used without exchange groups for gel chromatography (p. 71), ordinary adsorption chromatography – including high efficiency systems – and gas chromatography.

A term which is used in connection with adsorbents, usually without adequate definition, is 'activity'. It is not, in fact, an easy term to define in this context. To the surface chemist it usually relates to the specific surface of the solid; that is, the surface area measured in square metres per gram, in which case carbon, silica gel, and alumina can be made the most active of solids possessing specific surfaces of many hundreds of square metres. Others, like calcium carbonate and lime, have specific surfaces measured in tens of square metres and less, and can therefore be considered relatively inactive. This definition of activity relates to the *amount* of substance adsorbed at a given concentration and says nothing about the tenacity with which it is held. On the other hand, the term 'activity' is often used to denote the *strength* of adsorption and this is usually the sense referred to in chromatography and the one that will be employed here. The term 'chromatographic activity' will be used. (In a narrower sense still, the surface chemist may regard activity in terms of catalytic power in heterogeneous catalysis). Often, of course, the two types of activity are found together, that is, a solid with a large surface area adsorbs tenaciously, but the situation where a substance is firmly held on the surface of a 'low area' solid is not uncommon. Thus it might be expected that substances possessing some acidic character should be strongly adsorbed on say, lime or magnesia. As might be expected, chromatographic activity is the more specific and it is found that the strength of adsorption of polar groups on polar compounds increases in the order:

$$-CH=CH- < -OCH_3 < -CO_2 R < =C=O$$
$$< -CHO < -SH < -NH_2 < -OH < -CO_2 H$$

this order is approximately reversed for carbon (see also under 'Solvents').

The shape of the isotherm associated with 'active' adsorption is illustrated in Fig. 2.6, curve I. Since it is more strongly curved than II it will give rise to more pronounced tailing, therefore the specificity mentioned above, though desirable, is often associated, in the case of the more polar compounds at least, with steep isotherms and hence pronounced tailing. Two undesirable chromatographic features may arise as consequences; first, very slow movement of the

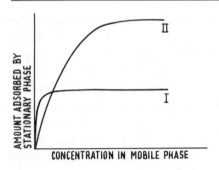

Fig. 2.6. Adsorption isotherms

adsorbate on the column and, second, wide chromatographic bands of low concentration with a tendency to overlap. For these reasons it is often desirable to deactivate the adsorbent or to use the stepwise or gradient elution techniques mentioned above. Deactivation may be effected by heating the solid to a high temperature, for example, 700–800°C in the case of alumina, but it is more usual to treat the solid with a very strongly adsorbed substance such as water before use on the column. In either case the effect is to block the more active sites or close the fine pores of the adsorbent, with reduction in the specific surface and a tendency to straightening of the isotherm. This aspect is also considered in some detail in Chapter 4 (p. 243). When large amounts of water are used for deactivation it is sometimes difficult to decide whether in fact the mechanism of separation has changed from adsorption to partition.

Preparation of adsorbents

Many adsorbents such as alumina, silica gel, active carbon, and magnesium silicate can be obtained commercially. They often require activation before use; this can be achieved by heating, possibly in a vacuum, when the adsorbent loses water and other adsorbed materials. There is usually an optimum temperature for activation and for alumina, it is about 400°C. The period of heating is also important because prolonged baking will cause a loss of activity which may, or may not, be desirable. Three or four hours is usually sufficient. In the absence of further information, heating at 200°C for four hours is safe for most solids.

If the freshly activated solid proves too active for the required separation, deactivation may be carried out by the controlled addition of water, as already mentioned. The subsequent activity is related to the amount of water added. Thus in the activity scale used by Brockmann and Schodder[9] for alumina, Grade I is the most active and is simply alumina activated at about 350°C for several hours. Grade II has about 2–3% of water added, Grade III 5–7%, Grade IV 9–11%, and Grade V (least active) about 15%. The activity is tested by the relative adsorbability of a number of azodyes which are, azobenzene, *p*-methoxyazobenzene, Sudan Yellow, Sudan III, *p*-phenylazoaniline, and *p*-phenylazophenol. These dyes are used in pairs on chromatographic columns to determine the grade of the adsorbent, and are listed in the order of their adsorbability, that is, azobenzene is the least strongly adsorbed and *p*-phenylazophenol the most strongly adsorbed. Grade I is the only one which adsorbs azobenzene, Grade II does not retard it at all, but retains *p*-methoxyazobenzene. Grade III does not retard *p*-methoxyazobenzene significantly but does Sudan Yellow, and so on. Hernandez *et al.* [10] described procedures for activating and standardising silicic acid (silica gel) by three methods to give all the Brockmann activity grades, including two new grades of intermediate activity. Williams, Hagdahl and Tiselius[4] found that the adsorption isotherms of substances like saccharides and amino-acids on activated carbon were depressed (that is, the carbon was deactivated) when the adsorbent was treated with, for example, cetyl alcohol.

As may already be apparent, the principal disadvantage of

adsorbents, apart from their tendency to produce tailing, is the difficulty of preparing samples with reproducible properties. Even in the same batch adsorbents may vary. They also 'age', or de-activate on keeping. There are complicated instructions of the most empirical nature in the literature for the preparation of adsorbents for chromatography (see, for example, the method for preparing silica gel described in the next section and the Brockmann method described above), and until recently, with the renewed interest in liquid chromatography generated by high efficiency systems (p. 85), there has not been much systematic investigation except by manufacturers specialising in the production of adsorbents. Sing and Madeley[11] studied, by means of nitrogen adsorption, the preparation of silica gel from sodium silicate and sulphuric acid. They found that the final pH of the solution was most important; for example, the product obtained at pH 3.72 was quite different from that obtained at pH 5.76; at pH 3.72 the specific surface was 830 $m^2 g^{-1}$ and at pH 5.76 348 $m^2 g^{-1}$. Provided that the same pH is reached during preparations and the same drying procedure is followed, then the surface properties are remarkably constant from batch to batch. The batch size has some influence, but this effect can be eliminated[12] if an acetic acid/sodium acetate buffer is used. In this case only the final pH is important, especially if it is about 4.64, the maximum buffer capacity of the system.

The particle size of the adsorbent should be small enough to minimise diffusion effects and ensure that equilibria are rapidly attained; on the other hand, the particles should not be so small that they will impede the flow of solvent. The size fractions between 100 and 200 B.S.S. are usually suitable, although the more uniform the particle size the better since this leads to more even column packing. The main drawback in using closely sized fractions is the large amount of wastage during sieving; proprietary materials tend to be rather expensive on that account.

Reactions on columns

The active solids used for packing chromatographic columns are often good catalysts; this is an additional hazard which must be taken into account. Alumina, especially when alkaline, is particularly liable to bring about chemical change and many reactions on columns have been reported. Alkaline alumina can cause the condensation of aldehydes and ketones, and when such reactions are liable to occur a neutral alumina should be used (suitable alumina is commercially available). Neutralisation can be effected by boiling with water, or first with dilute nitric acid, followed by water until neutral. Acetone is not recommended as a solvent for use with alumina columns. Some irreversible adsorption may take place on columns, and for this reason complete recovery of the adsorbates is not always achieved. Isomerisation of various compounds such as terpenes and sterols has been reported to occur on silica gel. Charcoal is particularly prone to giving adsorption losses and can also bring about chemical decompositions.

Dry column chromatography

The system of column chromatography so far described has been one in which the adsorbent is in contact with the mobile phase from the time that the column is made up. Once the adsorbent has been made into a slurry with the solvent in readiness for column packing, liquid and solid are kept in contact continuously thereafter — hence the warning on p. 21 that columns should not be allowed to run dry.

In the technique of thin-layer chromatography (TLC, Chapter 5), an adsorbent is spread as a thin film on a glass plate, and dried, and the solvent is caused to flow through the dry adsorbent, carrying the constituents of the sample through different distances along the plate. This method is a very small-scale one, and it gives better resolution than can usually be obtained on conventional columns; it is also quicker. Dry column chromatography is a method in which the high resolution and speed of TLC are obtainable in column chromatography on a preparative scale[115, 116].

The column. The adsorbent is poured as a fine dry powder into the column container. Glass columns with a support for the adsorbent as shown in Fig. 2.1 may be used. The tap is left open while the powder is poured in, and the column is tapped regularly during filling, or a vibrator may be held against it. This way of filling a column is most commonly used to pack the long narrow columns used in high-efficiency liquid chromatography (p. 90) and gas chromatography (Chapter 4).

An alternative type of column for dry column chromatography is a tube 2.5–7.5 cm in diameter made from nylon film about 1.6 μm thick. Suitable tubing can be bought in rolls. An appropriate length is cut off and sealed at one end by heating or stapling, and a small plug of glass wool is inserted into the closed end. A few holes are pierced with a needle below the glass wool, to allow the escape of air when the column is being filled. The dry adsorbent is poured in slowly, with shaking or vibrating. The completed column is rigid enough to be held in an ordinary clamp.

Adsorbent and mobile phase. Alumina has been used most commonly as the adsorbent, although the use of silica gel has been reported (see TLC, p. 283). Alumina may have to be less active than alumina used for a similar separation in a normal column. The solvent systems are similar to those already described.

Using the column. The sample, in a small volume of mobile phase, is allowed to run slowly from a pipette on to the top of the column; it must sink completely into the adsorbent before the flow of mobile phase is begun. An alternative method of loading is to dissolve the sample mixture in a volatile solvent, to mix the solution with a few grams of adsorbent, and to allow the solvent to evaporate. The adsorbent, now carrying the sample, is then spread evenly over the top of the column and covered with a layer of sand or glass beads. The flow of mobile phase is started by running it carefully on to the top of the column so that a layer about 5 cm in depth stands above the solid surface. The liquid flows down the column over the surface of the adsorbent particles, with a visible moving solvent front. When the front reaches the bottom of the column development is complete. The separated substances can be detected on the column by standard methods as described on p. 22.

Partition columns

In the systems described in the previous section the process of separation depends on the use of a solid which adsorbs the various substances to be separated as the solution containing them passes over it. Differences in the affinity of the components of the mixture for the solid surface and solubility in the flowing solvent result in the resolution of the mixture. In partition chromatography the solid adsorbent is replaced by a stationary liquid which is normally only partly miscible with the flowing liquid. A solute will distribute itself between the two liquid phases (stationary and mobile) according to its partition coefficient. The basis then of this method is that, due to differences in the partition co-efficients of the various components, the mixture will be resolved in much the same way as in the adsorption systems. Partition chromatography was introduced by Martin and Synge[16] in 1941.

The stationary phase must be supported in some way. It would in fact be possible to dispense with the support by allowing one solvent to flow dropwise through the other, but this, although common as an extraction technique, does not give a separation in the chromatographic sense. It is, however, the basis of the Craig counter-current technique which amounts to a series of conventional two-phase solvent extractions carried out in succession, usually in a partly or fully automatic machine. The theory of this method is discussed in Chapter 4.

For chromatographic purposes the stationary liquid is supported on a solid which, as nearly as possible, is inert to the substances to be separated. The coated solid is packed in columns as in adsorption chromatography. There is, in fact, very little visible difference between the two types of column. The support material must adsorb and retain the stationary phase, and must expose as large a surface of it as possible to the flowing phase. It must be mechanically stable and easy to pack into the column when loaded with the stationary liquid, and it must not impede the solvent flow.

Needless to say there is no support which has all these properties to the desired extent. The greatest difficulty (more or less unavoidable when a solid has to be used) is the incursion of adsorption effects. Even if the surface of the support is completely covered with liquid, adsorption effects can still make themselves felt. As complete coverage of the surface is not easy to achieve adsorption may be a major influence on the separation. As far as liquid-phase chromatograph on columns is concerned it is probably true to say that the division into adsorption and partition methods is of practical, rather than theoretical, significance. The Craig process may well be the nearest approach to 'pure' liquid-liquid chromatography. The importance of adsorption varies from system to system and is mentioned briefly in connection with the different supports described below.

The moving phase in partition chromatography may be a liquid or a gas, and the general principles are the same in each case. The theory is dealt with in Chapter 4 and only the practical aspects of liquid-liquid partition will be referred to here.

Solid supports

The supports most commonly used are silica gel (sometimes referred to as silicic acid), cellulose powder, and diatomaceous earths (kieselguhr, Celite, etc.). Other solids such as starch or glass beads have found more limited use.

Silica gel is almost always used with water or a buffered aqueous solution as the stationary phase. The amount of liquid is about 0.6 cm^3 g^{-1} of gel. It is added carefully so as to obtain as homogeneous a mix as possible. A typical method for the preparation of the gel itself is to mix one part of sodium silicate (water-glass) with one part of water and one of ice. Concentrated hydrochloric acid is then added with vigorous stirring until the mixture is strongly acid. After filtering, the precipitated gel is allowed to stand for two or three days in 3% hydrochloric acid. It is then freed from iron by boiling with 3% hydrochloric acid, washed free from chloride and dried at 80°C.

As already mentioned in the previous section, the surface properties of silica gel vary according to the pH at which it is precipitated. No comparisons of the surface properties with its effectiveness as a support for partition chromatography seem to have been made, but the fact that differences have been observed draws attention to the very empirical nature of most of the chromatographic work reported with this system. It is fairly certain that adsorption plays a large part in all separations employing silica gel, but it is, nevertheless, extensively used. The same kinds of chemical reaction can occur in partition systems as in adsorption systems, although they may be less marked due to the diminished influence of the surface of the solid support caused by the stationary liquid. Silica gel used for chromatography is in the form of a fine white powder; a fairly narrow range of particle sizes is desirable – about the same as for adsorption columns –

although swelling may occur when the gel is mixed with the stationary phase.

Diatomaceous earths; these are all similar, being available commercially as kieselguhr, Celite, or other proprietary products. The amount of liquid phase used with these solids is about 0.8 cm^3 g^{-1}. They are usually pure enough for use, but if necessary they can be freed from iron in the same way as silica gel.

Cellulose powder is supplied ready for use and usually requires no further treatment, not even the addition of the stationary phase, since this is acquired from the aqueous solvent. The use of cellulose in columns is an alternative to the use of cellulose in the form of paper sheets (Chapter 3) or in thin layers coated on glass plates (Chapter 5). The main reason for using columns instead of sheets or thin layers is that in effect a third dimension is added and hence larger quantities of materials can be separated. Cellulose columns are essential if a preparative separation is required, and they have also been found more convenient for quantitative estimations. The separations achieved on columns and thin layers are similar to those on paper, but are not necessarily identical, since the cellulose fibres have some sort of regular orientation in paper sheets, whereas in cellulose powder they have a completely random arrangement. Another difference, which, like the first, may affect the solvent flow, is that in a column the support and stationary phase are in contact with mobile phase before the separation starts, whereas on paper and thin layers the mobile phase has a definite boundary which moves ahead of the solutes.

While the separation on cellulose is mainly due to partition, adsorption again plays some part and ion-exchange is also possible. The extent of adsorption is uncertain, but it is partly due to the polar nature of the hydroxyl groups of the cellulose molecule, and varies according to the polarity of the solutes. The ion-exchange effects are also due to the hydroxyl groups and to the small number of carboxyl groups in the cellulose. These groups act as weak acid ion-exchangers, the protons exchanging with cations in the solution. The same factors influence separations on paper. Apart from the methods of handling, the main difference between the use of cellulose as powder in columns and as sheets of paper or thin layers is that in the latter two cases the separated substances are detected and identified in their final positions on the sheet or layer, whereas in the column method the substances are normally eluted and identified in the eluate. Cellulose powder should be stored and treated in the same way as paper sheets (Chapter 3).

Apparatus

The columns and other apparatus used for partition chromatography are much the same as those used for the adsorption technique. Tubes for most laboratory scale separations are between 25 and 50 cm in length, and up to 4 cm in diameter. Fairly long narrow columns are favoured in preference to short wide ones, although it is not possible to generalise about the exact ratio of diameter to length required. Since it is not usual to extrude partition columns, a glass wool plug can conveniently be used to support the packing. The flow of solvent through partition columns tends to be rather slow, but can be accelerated by the application of air pressure at the top as already mentioned for adsorption columns (p. 23).

Preparation of the column

Careful packing of the column is just as important as in adsorption chromatography. The dry-looking mixture of support plus stationary phase is stirred with some of the mobile phase to be used initially. The slurry is then poured in small portions into the tube which contains a little of the same liquid. Each portion is tamped down with a 'rammer', which may be a long glass rod with a flattened end or, better, a perforated metal disc, a little smaller in diameter than the tube and attached to a long handle. Each portion of solid must be first thoroughly stirred up and then packed down with firm slow strokes; successive portions being well mixed with the top of the preceding portion to avoid striation. Excess of solvent may be allowed to run through the column or removed with a pipette. For good results the procedure must be carried out with some deliberation; the success of a separation depends on the formation of compact and regular chromatographic bands. As in adsorption chromatography some practice may be necessary before complete success is achieved.

Application of the sample

One method of applying the sample is to use a pipette with a bent tip in the way that has been described for adsorption columns. Another method is to use one or two small discs of filter paper (perforated with very small holes) of a size just to fit on top of the column. The mixture to be separated is dissolved in a suitable solvent (volatile) and adsorbed on one of the discs, the second being used if necessary for any washings (in quantitative work). After drying, the discs are placed on top of the column (just run dry), covered with a very thin layer of the solid support, and subsequently moistened with a few drops of the solvent. A third method is to dissolve the mixture in some of the mobile phase and to mix it with some of the supporting solid until a dry powder is obtained. The powder is then put on top of the column, already just run dry, and a little of the mobile phase is added. This method is very convenient when cellulose is being used as the support.

Elution

The elution technique used with partition columns is exactly similar to that used on adsorption columns. Most of the separations reported in the literature have used an eluting solvent of unvarying composition, but gradient elution may sometimes give better results. The other techniques, such as frontal analysis, are not suitable for partition chromatography.

If the stationary phase is aqueous, the mobile phase is an organic liquid or mixture. Some typical separations are shown in Table 2.2. If a hydrophobic support is used, an organic stationary phase and an aqueous mobile phase become possible. This is known as 'reversed phase' chromatography, and has been used to separate substances which are too soluble in organic solvents to give a good separation with the conventional systems.

Identification of separated substances

Substances can be identified in the same way as in adsorption chromatography, either by the collection of suitable fractions and subsequent examination, or by the continuous monitoring of the eluate – see page 84.

Ion-exchange chromatography

Ion-exchange chromatography on columns is confined almost exclusively to the use of ion-exchange resins, mainly because of the desirable properties of these materials, such as

Table 2.2 *Some typical separations on partition columns*

Separation	Support	Stationary phase	Mobile phase	Ref.
C_1-C_4 alcohols	Celite	water	$CHCl_3$ or CCl_4	13
C_2-C_8 fatty acids	silica gel	water (buffered)	$CHCl_3$/BuOH	14
C_1-C_2 fatty acids	silica gel	water	Skellysolve/Bu_2O	15
acetylated amino-acids	silica gel	water	$CHCl_3$/BuOH	16
acetylated amino-acids	kieselguhr	water	$CHCl_3$/BuOH	18
amino-acids	starch	water	PrOH or BuOH/HCl	17
proteins (ribonuclease)	kieselguhr	water	$(NH_4)_2SO_4$/H_2O/cellosolve	19
purines	starch	water	PrOH/HCl	20
17-oxo-steroid glucuronides	silica gel	aqueous sodium acetate	$CHCl_3$/EtOH/AcOH	21
corticosteroids	Celite	water	EtOH/CH_2Cl_2* 40−60 petrol/CH_2Cl_2*	22,23
methoxy aromatic acids	silica gel	0.25 mol dm^{-3} H_2SO_4	BuOH/$CHCl_3$	24
phenols	cellulose	water	MeOH/BuOH/$CHCl_3$	25
DNP amino-acids	chlorinated rubber	butanol	aqueous buffers	26
17-oxo-steroids	silica gel	water	CH_2Cl_2/petrol*	26
inorganic	cellulose	water	acetone/HCl	27
dibasic acids	silica gel	water	BuOH/$CHCl_3$ (stepwise in three mixtures)	28
alkanes and cycloalkanes	silica gel	aniline	iso-PrOH/benzene	29
organic acids	silica gel	aqueous sulphuric acid	$CHCl_3$/BuOH*	30
lanthanides	kieselguhr	tributyl phosphate	(i) 15.8 mol dm^{-3} HNO_3† (ii) 15.1 mol dm^{-3} HNO_3 (iii) 11.5 mol dm^{-3} HNO_3	31
lipids	silica gel	water	various	32

* Gradient elution
† Stepwise elution

mechanical and chemical stability and the uniformity of their bead (particle) sizes. Cellulose powder chemically modified to contain ion-exchange groups is also used for separations on columns. Sheets of the same treated cellulose, and cellulose sheets impregnated with ion-exchange resins[33], can be used, with the techniques of paper chromatography, for separations involving ion-exchange (p. 158). Sheets are also impregnated with substances such as zirconium phosphate[34] and ammonium molybdophosphate[35]. Many of these materials are also conveniently used in the form of thin layers spread on suitable inert supports (see Thin-layer Chromatography, Chapter 5).

There are many books and reviews dealing with the manufacture, characteristics, and uses of ion-exchange resins, since these materials are used in large quantities for the complete removal of ions from solution, both on the laboratory and industrial plant scales. Complete de-ionisation of solutions is a non-selective use and will not be considered further here; only separations of the chromatographic type where the resin takes the place of the adsorbent in adsorption chromatography will be described. Some spectacular separations have been achieved, notably of the lanthanides, actinides, and amino-acids.

Synthetic ion-exchange resins in common use are based upon an insoluble matrix of a high polymer, which is usually polystyrene, but there are also some based on poly-methacrylic acid. The former type is made by the polymerisation of styrene in the presence of a small amount of divinyl benzene, which gives a controlled amount of cross-linking — an important factor in chromatography. The cross-linking renders the polymer insoluble; too little makes the resin liable to excessive liquid uptake and hence pronounced swelling, whereas too much reduces the exchange capacity, probably due to steric hindrance. The polar groups which confer the ion-exchange properties are introduced after polymerisation except for polymethacrylic acid. By conducting the polymerisation in an aqueous emulsion, beads of definite sizes can be produced, and it is in this form that the resins are used for de-ionising and chromatographic purposes. Some resins are made in sheet form giving ion-exchange membranes. They have no use in chromatography as such, but may be used for 'desalting' solutions, which may be an essential preliminary to an attempted chromatographic separation. In Chapter 3 a brief description is given of the use of these membranes.

The two fundamental types of ion-exchange material are cation- and anion-exchangers, which in turn may be subdivided according to their strengths as acids or bases. The polar groups in cation-exchangers are acidic, either $-SO_3H$ or $-CO_2H$. They are attached to the polymer molecule in a regular way, and are accessible to the solution containing the ions to be removed or separated. The polar groups in anion-exchangers are tertiary or quarternary ammonium groups ($-CH_2 \cdot NR_2$ or $-CH_2 \cdot NR_3^+$) and they function in an analogous manner. Table 2.3 illustrates the general characteristics of the various resins. Anion-exchangers are usually supplied in the chloride form rather than as the hydroxide because of its greater stability.

To illustrate how different cations may be separated on a column packed with a strong-acid resin one may consider a single resin bead in the hydrogen form. When the bead is immersed in water the sulphonic groups are fully ionised, but the protons, although more or less free to wander within the resin matrix, cannot escape from it because of the demands of electroneutrality and the absence of anions in the surrounding liquid. When another cation in solution comes close to the bead, however, exchange between the resin-bound proton and the cation in solution becomes possible because the solution must contain anions as well. On the macro-scale the equilibrium is determined by the Law of Mass Action; in the exchange reaction:

$$M^+ + R-H \rightleftharpoons H^+ + R-M$$

M^+ is the free cation and $R-H$ symbolises the resin-bound proton, hence using activities, one can write the equation:

$$K_d = \frac{a_{H+} \, a_{R-M}}{a_{M+} \, a_{R-H}}$$

where K_d is sometimes known as the 'equilibrium distribution coefficient' or as the 'selectivity coefficient'. For dilute solutions activities may be replaced by concentrations; in any case there is no entirely satisfactory way of measuring the activities of the ions on the resin. As a solution containing M^+ ions flows through a column of resin in the hydrogen form the M^+ will be replaced by H^+ ions according to the value of K_d.

Table 2.3 *Ion-exchange resins*

	Type	Exchanging group	Effective in the pH range	Exchange capacity*
Cation-exchange	Strong acid	$-SO_3H$	1–14	4 mmol H^+ g^{-1}
	Weak acid	$-CO_2H$	5–14	9–10 mmol H^+ g^{-1}
Anion-exchange	Strong base	$-CH_2 \cdot NR_3^+$	1–15	4 mmol OH^- g^{-1}
	Weak base	$-CH_2 \cdot NR_2$	1–9	4 mmol OH^- g^{-1}

$$R = -CH_3 \text{ usually}$$

* The capacity quoted is for the dry resin

If only a small amount of the solution is used and it is washed down the column with pure water, all the M^+ ions will eventually be replaced by hydrogen ions, and will form a stationary adsorbed band. The distribution of the ions within this band will depend on the value of K_d; if large, the band will be narrow and concentrated; if small, wide and diffuse. Thus if a few cm^3 of 0.1 mol dm^{-3} sodium chloride are placed on a column which is then washed with distilled water, the sodium ions will remain in a more or less narrow band near the top of the column and an equivalent amount of hydrochloric acid will be liberated to be eluted from the column. (In fact it is possible to standardise a solution of caustic soda by using it to titrate the hydrochloric acid eluted when a solution of a known weight of sodium chloride is passed through a column in the hydrogen form.) In order to make the adsorbed band of ions (M$^+$) move down the column, water is ineffective and it is obviously necessary to elute with a solution of an acid (or a solution containing another cation) so that exchange can take place. The M$^+$ ions will then be washed out of the column, leaving it in its original form. The rate at which the band of ions moves will depend on the pH of the eluting acid and the value of K_d. Thus, two ions having different affinities for the resin will move at different rates down the column and a separation will be achieved.

Selectivity of resins

Nature of the resin. The selectivity coefficient (K_d) is influenced by a number of factors operating in the resin; most obvious is the acid or base strength, or in other words, the nature of the polar groups. Since in cation-exchangers the exchange groups are either $-SO_3H$ or $-CO_2H$ there is little scope for differences within each class (strong or weak). Anion-exchange resins, however, allow of more variation because the exchange groups are either $-NR_3^+$ or $-NR_2$ and the nature of R can be changed. The ionic form of a resin will also have some influence; the nitrate form of an anion resin will behave slightly differently (though not markedly so) from the chloride form; a series of K_d values is valid for only one specific ionic form of the resin. The selectivity can also be affected by the degree of cross-linking. Small amounts reduce the selectivity while large amounts may totally or partly exclude larger ions – particularly organic ones. Selectivity is also slightly influenced by temperature.

Nature of the exchanging ion. The strength with which ions with different charges are held decreases in the following way:

$$M^{4+} > M^{3+} > M^{2+} > M^+,$$

an ion such as Th^{4+} being more firmly held than K^+. This order is however, only true for solutions whose strength is about 0.1 mol dm^{-3} in the exchanging ion. At higher solution strengths the tendency is for the M^+ ion to be most firmly bound, the others remaining in the same relative order. Within a particular series of ions carrying the same charge there is also a range of selectivity, for example for Dowex 50[37] the order is:

$$Ba^{2+} > Sr^{2+} > Ca^{2+} > Mg^{2+} > Be^{2+}$$
$$Ag^+ > Tl^+ > Cs^+ > Rb^+ > NH_4^+ > K^+ > Na^+ > H^+ > Li^+$$

and for Dowex 1, the following series holds:

$$NCS^- > I^- > NO_3^- > Br^- > CN^- > HSO_4^- \equiv HSO_3^-$$
$$> NO_2^- > Cl^- > HCO_3^- > CH_3CO_2^- > OH^- > F^-$$

For Dowex 2 the OH^- ion lies between the Cl^- and HCO_3^-. As a rough guide; the affinity decreases as the radius of the *hydrated* ion increases. Strelow[36] published a table of K_d values for 43 cations on sulphonated polystyrene. For organic compounds such as amino-acids similar orders apply, with some disturbance due to adsorption and ionic size. The latter effect is particularly important in organic compounds and the degree of cross-linking may be critical. Brimley and Barrett[117] gave a useful review of the order of displacement of amino-acids from different resins.

There is little effect on the selectivity by the non-exchanging ion unless complex formation is possible. That aspect is considered in more detail under 'Separation methods'.

Properties desirable in resins

For most applications of ion-exchange resins an important factor is the accessibility of the exchange sites to the ions in solution. The exchange takes place partly in the thin film of solvent adsorbed on the surface of the beads, and partly within the resin matrix; it is normally assumed that for small ions (including all metallic ions and simple inorganic anions, and many small organic ions) all the sites are equally accessible to the displacing ions in solution. This accessibility within the lattice depends partly on the degree of cross-linking of the polymer chains, and if that varies in different parts of the resin, the exchange properties will be variable also, and that is not desirable for chromatography. With small amounts of cross-linkage exchange equilibria are established more quickly, due to the extra swelling of the resin, so that the diffusion of ions becomes more rapid. The degree of cross-linking is controlled in the manufacturing process; a larger range of cross-linking can be achieved in the poly-styrene/sulphonic acid resins than in polymethacrylate or basic resins. For chromatography a resin must also possess the following properties:

1. It must possess mono-functional exchange groups. There is no difficulty about this with modern resins, but

earlier products made from phenol were polyfunctional ($-OH$ and $-CO_2H$ groups) and their exchange properties depended on the pH of the solution in which they were immersed. They were not, on this account, suitable for chromatography.

2. It must have a controlled degree of cross-linking; 4–8% is best for chromatography.
3. The range of particle sizes must be as small as possible.
4. The particle size must be as small as practicable.

Particle size

For large-scale (industrial plant) operation a fairly high flow rate through the resin bed is desirable, and that requires the use of fairly large particles. The standard size for analytical purposes is about 50 mesh B.S.S. Analytical grades of resin differ from the standard industrial grades in being washed free of 'fines' and water soluble traces of polymerisation intermediates. The useful bead sizes for chromatography are:

1. 400–600 mesh   Only two types of resin, both cation-exchangers, are obtainable in this size.
2. 200–400 mesh   Lower limit of practicable size for manufacture of anion-exchange and some cation-exchange resins. Even with this size flowrates are slow, but useful for microscale work.
3. 100–200 mesh   Macro-scale separations; quantitative.
4. 50–100 mesh   Preparative separations.
5. 14– 50 mesh   Industrial scale separations.

In table 2.4 the products of two principal manufacturers of resin are listed, only the resins supplied in chromatographic grades being shown. There are other products with similar properties and the list is not intended to be exhaustive.

Table 2.4 *Some commercially available resins suitable for chromatography*

Name	Type	Functional group	Bead sizes	% cross-linking	Exchange capacity g^{-1} (dry resin)	Form supplied	Working pH range	
Zeo-Karb 225	Strong acid	$-SO_3H$	14–52, 52–100, 100–200, >200*	4, 8, 12, 20	4.5–5.0 mmol H^+	Na^+	1–14	
Amberlite CG 120	Strong acid	$-SO_3H$	100–200, 200–400, 400–600†	8	5.0 mmol H^+	Na^+	1–14	
Zeo-Karb 226	Weak acid	$-CO_2H$	14–52, 52–100*	2.5, 4.5	9–10 mmol H^+	H^+	6–9	
Amberlite CG 50	Weak acid	$-CO_2H$	100–200, 200–400, 400–600†		10.0 mmol H^+	H^+	5–14	
Deacidite FFIP	Strong base	$-CH_2\overset{+}{N}R_3$	14–52, 52–100*	2–3, 3–5, 7–9	4.0 mmol OH^-	Cl^-	1–14	R = alkyl
Amberlite CG 400	Strong base	$-CH_2\overset{+}{N}R_3$	100–200,† 200–400	8	3.8 mmol OH^-	Cl^-	0–12	R = alkyl
Amberlite CG 45	Weak base	$-CH_2NR_2$	100–200† 200–400		5.0 mmol OH^-	OH^-	0–9	R = alkyl

Ion-exchange cellulose materials are listed in Table 3.3 (page 160).

* British Standard Screens.
† U.S. Standard Screens.

Apparatus

For small-scale separations an ordinary burette, with a plug of glass wool on which to pack the resin, can be used.

A type of column commonly used in ion-exchange is illustrated in Fig. 2.7 (*a*). The resin is retained by a sintered or perforated glass disc, or a glass wool plug. The perforated disc is only suitable for the resins with the larger bead sizes. The side-arm shown in the figure prevents the top of the resin running dry, the column being filled so that its top is below the top of the outlet tube. The arrangement is suitable for desalting, but is not so useful for chromatography, as there may be a good deal of mixing in the outlet tube, leading to indistinct fractions or even the merging of successive fractions. A better arrangement serving the same purpose but preventing mixing is shown in Fig. 2.7 (*b*).

A better chromatographic separation can sometimes be obtained if the solution flows through a series of columns of diminishing size. Assemblies of this type are shown in Fig. 2.7 (*c*) and (*d*), (*c*) being distinguished by the use of

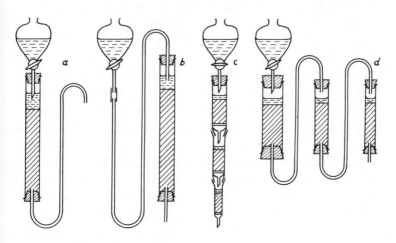

Fig. 2.7. Ion-exchange columns

ground-glass joints and sintered-glass discs. The multiple-column arrangements give sharper bands and more compact fractions. During its passage down the column a band of a particular ion tends to lose its original compact and level form. It diffuses and loses sharpness. A certain length of column is needed to give a separation and, if only one column of the required length is used, the zones may emerge in a very untidy condition. If, however, a zone is passed on to a second column it is adsorbed at the top and thus straightened. It then starts its progress down the second column with a compact form and level front. Three columns in series have been found to be sufficient in most cases; they become successively smaller because the loading of the later columns at any time is only a part of the original load.

Packing the column

When fresh resin is used it should be thoroughly washed by decantation until the supernatant liquid is clear; the slurry is then decanted into a column half full of distilled water and the resin allowed to settle. As with columns of adsorbent, the top of the resin must not be allowed to run dry, nor must there be any channels or bubbles in the column. With the larger bead sizes it is a good idea to back-wash the column (run water up through it) as shown in Fig. 2.8, to stir up the resin and ensure a homogeneous packing, but this technique may not be very successful with wider tubes. If back-washing is used a tube about twice as long as the packed resin should be employed to cope with the extra volume of liquid. Anion-exchange resins are more difficult to handle than cation-exchangers because their density is close to that of water, and consequently the top of the column is very easily disturbed. Care should therefore be taken to protect the top of the column from unsettling influences. A perforated glass disc placed on the packing is probably best, but a filter paper

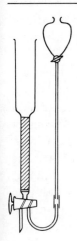

*Fig. 2.8. Arrangement
for back-washing*

cut accurately to the size of the column and pressed lightly
down will often suffice. When the eluting agent is being
dropped on to the small layer of liquid at the top of the
column, a low-density polythene disc floating in the surface
will minimise the agitation of the top layer of resin.

Separation methods

The strong acid and strong base resins find much wider
application than the weak varieties because of the wide pH
range over which they retain their exchange properties. The
weak acid and weak base types are highly ionised only when
in a salt form, which is why their operation is restricted to
the pH ranges indicated in Table 2.3. They may be preferred
for certain separations, however; polymethacrylic acid has
more than twice the exchange capacity of the sulphonic acids
at pH values greater than 7, and both weak acid and weak
base resins may show a greater selectivity in certain circumstances. A weak base resin will adsorb strong acids but not

weak ones and, conversely, weak acid resins will adsorb strong bases but not weak ones. Again, weak acid resins show greater selectivity for certain divalent ions.

Exchange isotherms measured in dilute solutions are approximately linear; that is, the amount of a particular ion held by a resin is directly proportional to the concentration of the ions in the solution in contact with it. At higher concentrations the uptake by the resin begins to fall off, with the result that the isotherm becomes concave to the concentration axis in a manner similar to that depicted in Fig. 2.6. Tailing can therefore be expected when the elution technique is used to separate concentrated solutions, otherwise symmetrical peaks, familiar in partition chromatography, will be obtained (see Experiment 6, Chapter 6). In chromatography the solutions used are mostly dilute; hence the elution technique is much employed and it frequently gives highly satisfactory separations. In addition, all the other methods described in Chapter 1 have been used, although frontal analysis is not often encountered. Displacement development and stepwise and gradient elution are all employed; reference to Table 2.5 will provide examples.

A modification of the elution technique, already referred to under 'Selectivity', is also used to good advantage. It depends on the alteration of the activities of the ions being separated by means of an eluting agent which will complex with them. A simple example is provided in Experiment 5 in Chapter 6 where copper(II) and iron(III) ions are separated by elution with phosphoric acid. If this separation is tried with hydrochloric acid as the eluant, the copper(II) ions tend to move more rapidly down the column than the iron(III) because they carry a smaller charge and are therefore not so firmly held by the resin. A good separation will not be obtained. If, however, phosphoric acid is used the iron(III) ions are very rapidly removed from the column, and a sharp separation results. Subsequently the copper(II) ions can be

Table 2.5 *Ion-exchange resins: inorganic separations*

Separation	Ion-exchanger	Method	Ref. No.
Lanthanides	Sulphonated polystyrene	Elution with citrate buffers	38
Lanthanides	Dowex 1-X10 in nitrate form − 325 mesh	Stepwise and gradient elution with $LiNO_3$ solution	39
Actinides	Dowex 50 in ammonium form	Elution with ammonium lactate or ammonium α-hydroxy-isobutyrate	40
Lanthanides from actinides	Dowex 1-X8	Elution using 10 mol dm^{-3} LiCl in 0.1 mol dm^{-3} HCl	41
Ca, Al, Fe	Dowex 2 in citrate form	Stepwise elution with (*a*) water, (*b*) conc. HCl, (*c*) dil. HCl. (Also involves backwashing with conc. HCl)	42
Ca, Sr, Ba, Ce	Dowex 50-X8	Stepwise elution with ammonium α-hydroxy-isobutyrate	43
Zr, Hf	Dowex 1-X8	Elution with 3.5% H_2SO_4	44
Zr, Ti, Nb, Ta, W, Mo	Dowex 1-X8 200−400 mesh	Stepwise elution with mixtures involving HCl, oxalic acid, H_2O_2, citric acid and ammonium citrate	45
Many simple mixtures of metal ions	Dowex 50W-X8 100−200 mesh (in polythene columns)	Stepwise elution with dil. HF followed by a mineral acid such as HCl or HNO_3	46

removed with hydrochloric acid. Clearly the phosphate ions form a much more stable complex with iron(III) ions, which are rendered colourless, than with copper(II). Complex formation is undoubtedly an important factor in other types of chromatography, particularly in inorganic separations on paper, but in no other technique has it been exploited to quite the same extent as in ion-exchange chromatography. One of the earliest and most spectacular successes of ion-exchange chromatography was the separation of the lanthanides on a strong acid resin with a buffered citrate solution for elution. Straightforward elution with hydrochloric acid brings about little separation, but the citrate ions complex with the M^{3+} ions, thus reducing their activity, which will now depend largely on the stability of the various citrate complexes. The equilibrium constants for the formation of the complexes (stability constants K_s) vary more than the affinities (K_d) of the free ions for the resin. Separation is therefore largely due to differences in K_s rather than in K_d. Similar operations have been carried out on the actinides with Dowex 50 in the ammonium form and elution with ammonium lactate or ammonium α-hydroxy-isobutyrate.

In Tables 2.5 and 2.6 some typical chromatographic separations performed on ion-exchange resins are listed.

'Salting-out' chromatography

In the technique of 'salting-out' chromatography ion-exchange resins are used for the separation of non-electrolytes by elution from columns with aqueous salt solutions. The method has been reviewed by Rieman[53] who has also done most of the work in this field. He found that a mixture of methanol, ethanol, and propanol was scarcely separated at all when eluted through a column of Dowex 1-X 8 with water, but when the experiment was repeated on the same column using 3 mol dm^{-3} ammonium sulphate solution as the eluant a good separation was obtained, the components appearing in the order of increasing molecular weight. In the presence of the salt the alcohols travel more slowly down the column because their solubilities are decreased and their affinities for the resin are increased. The solubility effect therefore considerably enhances the van der Waals adsorption, although the elution curves are symmetrical and show little sign of tailing. Other substances separated by this technique include ethers, aldehydes, ketones, and amines.

As far as resin type is concerned there appears to be little difference between Dowex-1 (anion) and Dowex-50 (cation), but the present of an exchange group (preferably strong)

Table 2.6 *Ion-exchange resins: organic separations*

Separation	Ion-exchanger	Method	Ref. No.
Amino-acids	Dowex 50-X4	Stepwise and gradient elution with citrate and citrate/acetate buffers	47
Phosphate esters	Dowex 1-X2 200–400 mesh in chloride or formate form	pH gradient elution (HCl)	48
Folic acid analogues	Diethylaminoethyl cellulose 100–250 mesh	Gradient elution 0.1–0.4 mol dm^{-3} phosphate	49
Chlorophenols	Dowex 2-X8 200–400 mesh in acetate form	Gradient elution: acetic acid in ethanol	50
Aminobenzoic acid, aminophenols, and related substances	Dowex 1-X10 200–400 mesh in chloride form	Stepwise elution: water; 1 mol dm^{-3}, 5 mol dm^{-3}, and 8 mol dm^{-3} HCl	51
Proteins	Diethylaminoethyl cellulose	Elution	52
Separation of acids from bases both of which are freely water soluble	Sulphonated polystyrene in H form	Stepwise elution: water to remove acids: dil. HCl for bases	

seems to be necessary in order to confer the correct physical properties on the resin. Smaller than normal exchange capacities do improve the resolving power of a resin, but reduction below about 4 mmol M$^+$g^{-1} reduces the rate at which equilibrium is established because the uptake of water by the resin is reduced; tailing may then occur.

Certain organic compounds of low solubility in water can be separated by a method which is almost the converse of the one just described. Instead of an eluant which depresses the solubility in water, one which increases the solubility is used. For example, Rieman separated the alcohols: t-amyl, amyl, hexyl, heptyl, octyl and nonyl on a column of Dowex 50-X 8 in the hydrogen form, with aqueous acetic acid as the eluant. The particle size of the resin was 200–400 mesh and the column dimensions 39.0 cm x 2.28 cm^2.

Inorganic ion-exchangers

It was mentioned at the beginning of the section on ion-exchange resins that certain inorganic salts had been used to impregnate paper to enable it to be used for separations involving ion-exchange. In recent years a considerable amount of work has been carried out to develop inorganic ion-exchange materials. Apart from clays and zeolites, which were in fact the first ion-exchange materials to be investigated, substances recently studied include heteropolyacid salts, zirconium salts, tin(IV) phosphate, tungsten hexacyanoferrate, and others. A general account of inorganic ion-exchangers was published by Amphlett[54] and chromatographic aspects were reviewed by Marshall and Nickless[55].

One of the reasons for the interest in inorganic materials is that resinous ion-exchangers are susceptible to radiation damage and are therefore not really suitable for use with highly active solutions, although they have been successfully used for the separation of, for example, mixtures of actinides. Inorganic materials possess other advantages including a much greater selectivity for certain ions such as rubidium and caesium, and the ability to withstand solutions at high temperatures. In addition, inorganic ion-exchangers do not swell appreciably when placed in water and there is no change in volume when the ionic strength of the solution in contact with them is changed. On the other hand, certain of the inorganic materials possess disadvantages such as solubility or peptisation at certain pH values at which resins are normally stable, or they may be soluble in solutions in which resins are insoluble. Zirconium molybdate, for example, dissolves in EDTA, oxalate, and citrate solutions. Again, they may exist in microcrystalline forms which are not very convenient for packing columns since they tend to impede the flow of mobile phase, although there are ways of overcoming this problem.

Inorganic ion-exchangers fall roughly into two groups,

crystalline, such as the ammonium salts of the 12-hetero-polyacids, and amorphous, such as zirconium phosphate. One example from each group will be described.

Of the crystalline variety, perhaps the most thoroughly investigated compound is ammonium molybdophosphate, $(NH_4)_3 PO_4 . 12MoO_3$ (AMP). As ordinarily prepared (see for example Thistlethwaite[56]) this salt is precipitated as very fine crystals which are unsuitable for packing columns. The problem was overcome by van R. Smit, Robb, and Jacobs[57], who used mixtures of AMP with asbestos. When equal weights of AMP and Gooch asbestos are mixed in water the AMP crystals adhere to the asbestos fibres. Columns packed with the coated asbestos thus obtained possess good flow characteristics. A method for the preparation of coarsely crystalline ammonium heteropolyacid salts including AMP was described by van R. Smit[58]; these crystals may be used to pack columns, again with good flow character-istics. Briefly, the process is to immerse large crystals of the heteropolyacid in saturated ammonium nitrate solution. The exchange capacity of AMP for caesium is about $1 \text{ mmol H}^+ \text{ g}^{-1}$ – that is, less than the theoretical value, but still comparable with the resinous ion-exchangers.

AMP can only be used in neutral or acid solutions because of its solubility in dilute alkalis. In acid solution (pH less than 2) only metal ions which form heteropolyacid salts insoluble in water exchange significantly with the ammonium ion, namely K^+, Rb^+, Cs^+, Ag^+, Hg_2^{2+} and Tl^+[57]. Those ions are therefore selectively adsorbed from solutions containing other ions and very sharp separations are possible. In addition, very good separations of, for example, sodium from potassium, potassium from rubidium, and rubidium from caesium have been carried out, and more complicated separations are undoubtedly possible[57]. Under acid condi-tions polyvalent ions do not exchange readily, but at pH 2–5, they are fairly strongly adsorbed, especially if a suitable buffer solution is used[59]. Simple group separations are possible, for example Sr^{2+} from Y^{3+}, a quite strong acid solution being necessary to elute the Y^{3+}.

Zirconium phosphate, which is easily prepared by the addition of a solution of zirconyl nitrate or chloride to a phosphate solution is extremely insoluble and will not dissolve in solutions of high or low pH. It is an inorganic polymer with a relative molecular mass of about 880[60]. It has a high exchange capacity – about 4 mmol OH^- g^{-1} in alkaline solution – and behaves as a weak acid cation-exchanger. It can be prepared in bead form[61] but does not show the very high selectivity of AMP. For example, the separation factor for caesium and rubidium on zirconium phosphate (α_{Rb}^{Cs}) is 1.3–1.5; in the case of AMP the corresponding figure is 26. Nevertheless zirconium phosphate shows adsorptive properties for a large number of cations and has been recommended for separations involving radioactive solutions.

Liquid ion-exchangers

There are a number of liquids, exemplified by such compounds as trioctylamine (TNOA) and bis-(2-ethylhexyl)-phosphoric acid (HDEHP) which are immiscible with aqueous solutions and yet possess the property of ion-exchange, the former with anions and the latter with cations. These liquids have been used for liquid-liquid extraction but they may also be used to impregnate solid supports such as cellulose powder and polytrifluorochloroethylene (Kel-F) powder which are then packed into columns in the same way as for partition chromatography. Such columns operate in much the same way as columns of ion-exchange resins and it is thus possible to combine the advantages of partition chromatography (reversed phase) with the selectivity of ion-exchange. The subject of liquid ion-exchangers has been comprehensively reviewed by Coleman, Blake, and Brown[62] and by Cerrai[63].

Early work was carried out with the liquid anion-exchangers only, their capacity for extracting cations from solution being dependent on the ability of the cation to form anion complexes such as $FeCl_6^{3-}$. Liquid cation-exchangers, mainly mono- and di-esters of phosphoric acid, were used later.

Coating the solid support. The method of coating the solid support is somewhat different from that used for the preparation of ordinary partition columns described earlier. If cellulose is to be used as the support it is first dried in an air oven at about 80°C and allowed to cool in a desiccator. The TNOA or HDEHP is dissolved in benzene or cyclohexane to make an approximately $0.1-0.3$ mol dm^{-3} solution and equilibrated with the aqueous eluant by vigorously shaking the solutions together. The cellulose is then immersed in the organic solution and vigorously shaken for a prolonged period. The excess of liquid is decanted and the wet powder almost completely dried with filter paper, any remaining solvent finally being removed in an air oven.

Kel-F powder may be coated in a similar way, although the preliminary drying is not necessary. Because of its highly porous nature Kel-F has a greater capacity for the liquid ion-exchanger than has cellulose. The method of packing the column is similar to that described for partition columns (p. 39).

Applications. A column in which TNOA is the stationary phase behaves in a similar fashion to a column packed with a strong anion-exchange resin, except that the capacity is somewhat lower; therefore similar separations may be carried out. Mostly, however, TNOA columns have been used for separating cations, usually as their chloride or nitrate complexes. Eluting agents therefore tend to be aqueous hydrochloric acid or nitric acid, the eluting power of such solutions increasing with decreasing concentration of acid. Cations of transition metals clearly lend themselves to this treatment and a number of separations have been described[63] on TNOA-cellulose columns, for example iron, cobalt, and nickel. Nickel is eluted first with 8 mol dm^{-3} HCl, cobalt is eluted next with 3 mol dm^{-3} HCl, and finally iron with 0.2 mol dm^{-3} HNO$_3$. Uranium, thorium, and zirconium have been similarly separated with 8 mol dm^{-3} HCl plus 5% conc. HNO$_3$.

HDEHP columns may be used to separate simple cations and it has been found possible to resolve mixtures of lanthanides, for example, lanthanum, cerium, neodymium, gadolinium, terbium, and thulium, by stepwise elution with hydrochloric acid. Lanthanum and cerium are eluted with 0.25 mol dm^{-3} HCl, neodymium, gadolinium, and terbium with 0.8 mol dm^{-3} HCl, and thulium with 6 mol dm^{-3} HCl. The eluting power of the acid solution increases with increasing strength in this case. It will be recalled that on a cation-exchange resin, little, if any, separation of lanthanides is possible with straightforward elution by hydrochloric acid and a complexing technique has to be employed.

It may be mentioned that paper chromatography with TNOA and HDEHP has been extremely successful. It is reversed phase chromatography and the technique is similar to that described on p. 157.

Gel chromatography

Adsorption studies on silica gel and active carbon had shown molecular sieve effects with materials of high relative molecular mass, and in 1954 Mould and Synge[64] showed that separations based on molecular sieving could be performed on uncharged substances during electro-osmotic migration through gels. This formed a basis for separations based on the relative sizes of molecules, and the systematic use of the principle was introduced in 1959 by Porath and Flodin[65], who used the term 'gel filtration' to describe their method of separating large molecules of biological origin in aqueous systems by means of polysaccharide gels. In a pioneering paper on non-biochemical uses Moore[67] used the term 'gel permeation chromatography' (GPC). Both of these terms are still used in their respective fields, and others have been proposed, but in 1964 Determann[66] suggested that 'gel chromatography' was the most general name for the technique, and this is the one that will be used here.

The stationary phase is a porous polymer matrix whose pores are completely filled with the solvent to be used as the mobile phase. The pore size is highly critical, since the basis of the separation is that molecules above a certain size are totally excluded from the pores, and the interior of the pores is accessible, partly or wholly, to smaller molecules.

The flow of mobile phase will cause larger molecules to pass through the column unhindered, without penetrating the gel matrix, whereas smaller molecules will be retarded according to their penetration of the gel. The principle is illustrated in Fig. 2.9.

The components of the mixture thus emerge from the column in order of relative molecular mass, the largest first. Any compounds which are completely excluded from the gel will not be separated from each other, and similarly, small molecules which completely penetrate the gel will not be separated from each other. Molecules of intermediate size will be retarded to a degree dependent on their penetration of the matrix. If the substances are of a similar chemical type they are eluted in order of relative molecular mass.

Adsorption effects on the surface of the gel particles can usually be ignored, and thus gel chromatography can be looked upon as a kind of partition chromatography. The

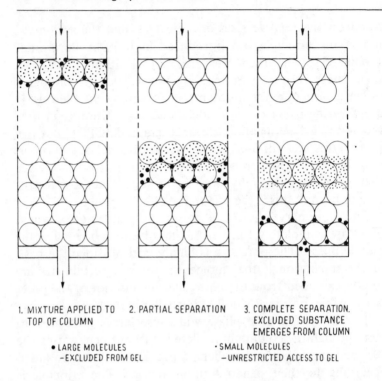

1. MIXTURE APPLIED TO   2. PARTIAL SEPARATION   3. COMPLETE SEPARATION.
   TOP OF COLUMN   EXCLUDED SUBSTANCE
                                                  EMERGES FROM COLUMN

● LARGE MOLECULES   · SMALL MOLECULES
  –EXCLUDED FROM GEL   – UNRESTRICTED ACCESS TO GEL

Fig. 2.9. Principle of gel chromatography

liquid stationary phase is the liquid within the gel matrix, and the mobile phase is the flowing eluant which fills the rest of the column. We have, in other words, a partition column where the two liquid phases, mobile and stationary, are of the same composition.

Gel chromatography was originally used for separation of biological materials, because the earliest gel media, cross-linked dextrans, were suitable for use only with an aqueous system. In 1964[67] cross-linked polystyrene gels suitable for use with organic solvents were first produced, and this made possible the extension of gel chromatography to the separation and characterisation of synthetic polymers. Application of the method, not only to a large variety of separations but also to the determination of relative molecular mass has proceeded rapidly in the last few years. Early accounts of the use of gel chromatography, were given by Porath[68], Tiselius, Porath, and Albertsson[69], Gelotte[70], and Granath[71]. References to more recent monographs on the subject will be found in the bibliography (p. 373). In addition, the manufacturers and suppliers of many commercially available gel media provide an extensive information service.

Column parameters and separations

A number of theoretical treatments of gel chromatography have been published, including applications of the theoretical plate concept similar to those outlined on pages 86 and 260. It is appropriate here to mention only a few simple parameters of most value in practical work.

A column is made up by pouring a slurry of swollen gel particles in the solvent used to swell the gel into a suitable tubular container. The total volume of the column, V_t (which can be measured) is the sum of the volume of liquid outside the gel matrix, V_o, the volume of liquid inside the matrix, V_i, and the volume of the gel matrix, V_m ; that is

$$V_t = V_o + V_i + V_m$$

V_o is also known as the void or dead volume; it is the volume of mobile phase which will elute a totally excluded molecule. The volume required to elute a particular molecule is the elution volume (or retention volume), V_e.

The use of these volumes is not very convenient in describing the behaviour of particular solutes because although they are characteristic of the gel and the solutes they also depend on the size of the column. The volumetric distribution coefficient, K_d, is more useful, as it is independent of column dimensions. It is defined as

$$K_d = \frac{V_e - V_o}{V_i}$$

K_d represents the fraction of the gel volume that is accessible to the molecule concerned, and is thus zero for a totally excluded molecule ($V_e = V_o$), and unity for a molecule which has access to the gel equal to that of the solvent ($V_e = V_o + V_i$). The relationship between these various parameters is shown in Fig. 2.10.

In practice, not all these volumes are readily determined. V_o can be measured as the volume required to elute a

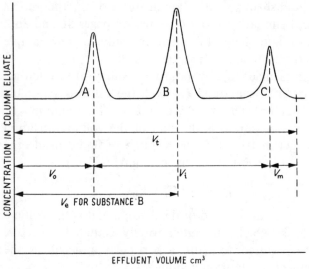

A – TOTALLY EXCLUDED MOLECULE
B – PARTLY EXCLUDED MOLECULE
C – MOLECULE WITH UNRESTRICTED ACCESS TO GEL PORES

Fig. 2.10. Separation parameters for gel columns

completely excluded solute, and V_e for a particular solute is equally readily measured. On the other hand the methods which can be used for determining V_i (requiring, for example, observation of the elution of tritiated water) give only approximate results. K_d is often, therefore, replaced by an alternative distribution coefficient, K_{av}, which is defined[72] as

$$K_{av} = \frac{V_e - V_o}{V_i + V_m}$$

that is,

$$K_{av} = \frac{V_e - V_o}{V_t - V_o}$$

Although all the volumes in this expression can be determined without difficulty, K_{av} has the disadvantage (because of the inclusion of a term for the gel volume, V_m) of not approaching unity for small molecules.

There is a linear relationship between K_d or K_{av} and $\log M_r$, and a plot of these values may be used to select a suitable gel for a separation (Figs. 2.11 and 2.12).

This plot shows the *fractionation range* of the gel, which is defined as the approximate range of M_r within which a separation can be expected, provided that the molecules concerned are in different parts of the range.

The upper end of the range ($K_d \to 0$) is the *exclusion limit*, the M_r of the smallest molecule which cannot penetrate the pores of the matrix.

The plot of K_d against $\log M_r$ diverges from linearity as K_d approaches 0 or 1, and the fractionation range is the range of M_r over the linear part of the curve. For most gels the lower limit of the range ($K_d \to 1$) is about one tenth of the exclusion limit.

Since M_r is not directly related to the size or shape of the molecule, it is necessary to state, in publishing fractionation

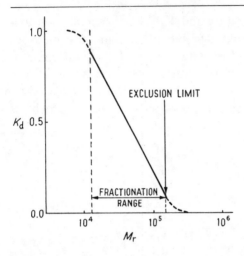

Fig. 2.11. Fractionation ranges

ranges for gels, what type of molecule has been used in the determinations. Most commercial gels are made in a number of grades of different fractionation range.

A gel is selected such that the M_r values of the substances to be separated lie on the straight part of the curve. This is illustrated in Fig. 2.12, where A (M_r c. 5×10^3) and B (M_r c. 6×10^4) are the substances to be separated. It can be seen that either of the gels II or III could be used. It would be better to use gel II because the K_d values are in that case further apart, and the fractions of the eluate containing A and B would be further apart also. More complex mixtures would require the use of gels of wider fractionation range, but in general it is best to use a gel of as narrow a range as possible.

Sorption effects by the gel matrix may alter the values of K_{av} and K_d for certain solutes, and the use of K_d against log M_r plots to select a gel takes into account neither those effects nor zone spreading in the column and variations in flowrate (which could arise, for example, from differences in

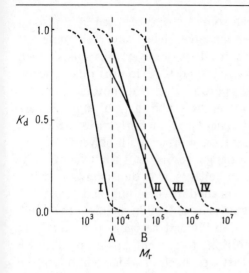

Fig. 2.12. Choice of gel

viscosity of sample and eluant). Taken together these may have an important effect on the separation, and hence on the choice of the best gel for a particular purpose.

Nature of the gel

The gel must be as chemically inert and as mechanically stable as possible. Gel materials are supplied in bead form, and as with ion-exchange resins a fairly uniform particle size is required, with a uniform porosity. Some examples of commercial gel materials are given in Table 2.7; the list is not exhaustive.

Two main types of gel material may be distinguished – xerogels and aerogels. Xerogels are gels in the classical sense; they consist of cross-linked polymers which swell in contact with the solvent to form a relatively soft porous medium, in which the pores are the spaces between the polymer chains in the matrix. If the liquid is removed the gel structure collapses, although it can sometimes be restored by replacing the liquid. Aerogels, on the other hand, are rigid materials which are not really gels at all; they are porous solids which are penetrated by the solvent, and they do not collapse when the solvent is removed; porous glass and porous silica are examples. Some gel materials, such as polystyrene, are xerogel-aerogel hybrids. These have a fairly rigid structure, but swell to some extent on contact with solvents, in the same way that ion-exchange resins do. An exception is agarose (see below), which behaves in an unusual way.

Much greater care is needed in handling and using the relatively soft xerogels and agarose than in handling the much more rigid polystyrene hybrids and aerogels, particularly in controlling column conditions to prevent particle breakdown and coagulation of the particles, both of which will retard the flow of mobile phase.

Dextran gels. The original gel chromatography medium, which is still probably the most widely used, was Sephadex G, a cross-linked dextran. Dextran is a natural linear polysaccharide (glucose-α-1,6-glucose) which is cross-linked by reaction with epichlorohydrin in dispersion in an organic solvent. The result is a water-insoluble solid in bead form. The material remains hydrophilic, however, and in water it swells to form a gel; the water regain is $1-20$ cm^3 g^{-1} of dry resin. Little swelling occurs in non-polar solvents, and thus dextran gels are only useful in aqueous media. Traces of carboxyl ($10-20$ mol H$^+$ g^{-1} of dry gel) remain, which may affect the separation of some polar species. Dextran ion-exchangers, in which ion-exchange groups are intentionally introduced, are described on page 78.

Dextran hydroxypropyl ether. Sephadex LH-20 is the hydroxypropyl ether of Sephadex G-25 (the number indicates the porosity of the gel). It will form a gel in both aqueous and organic solvents, and is particularly effective in solvents such as dimethyl sulphoxide, pyridine, and dimethyl formamide.

Polyacrylamide gels. Bio-Gel P is a polyacrylamide gel made by suspension co-polymerisation of acrylamide and N,N'-methylenebisacrylamide. (Lee and Schon[73]: Hjerten[74]). The pore size is regulated by variation of the proportions of the monomers. Polyacrylamide gels behave in a similar way to dextran gels, with a water regain of $1.5-18$ cm^3 g^{-1} of dry resin.

Polyacryloylmorpholine. Suspension co-polymerisation of acryloylmorpholine and N,N'-methylenebisacrylamide gives the medium known as Enzacryl Gel. This material swells to a gel in water, and also in pyridine and chloroform, although not in lower alcohols.

Polystyrene. Styrene-divinylbenzene polymers, as used in ordinary ion-exchange resins, but without ionising groups, will form gels in less polar organic solvents. One form of this polymer is Bio-Beads S. Unfortunately this type of polymer has a very small pore size, which rather restricts the range of uses.

If styrene and divinylbenzene are co-polymerised in a solvent mixture in which the polymer is sparingly soluble, a macro-porous form of the polymer bead is obtained. One such material is Styragel, which can be used very effectively in the gel chromatography of organic polymers in organic solvents. Introduction of a few $-SO_3H$ groups into a macro-porous resin (Aquapak) makes the gel hydrophilic, and thus usable in aqueous solvents, although the presence of the strongly acidic group may be a disadvantage.

Polyvinyl acetate. Co-polymerisation of vinyl acetate and 1,4-divinyloxybutane gives a material which forms a gel in polar organic solvents, including alcohols (Merck-O-Gel OR).

Agarose gels. All the xerogels described are characterised by extreme softness of the swollen gel when the pore size is large. Larger pore sizes with mechanical stability are obtainable in the aerogels glass and silica, but with a penalty of incursion of adsorption effects. To overcome this adsorption difficulty, and to obtain larger pore sizes, agarose gels were developed.

Agarose is a polysaccharide (alternating 1,3-linked β-D-galactose and 1,4-linked 3,6-anhydro-α-L-galactose) obtained from seaweed. Above 50°C it dissolves in water, and if the solution is cooled below 30°C a gel is formed, which is insoluble below 40°C. Above that temperature the gel 'melts' or collapses; freezing also causes irreversible changes in the structure of the gel. The chemical stability of agarose gels is similar to that of dextran gels.

Several forms of agarose are obtainable commercially (Sepharose, Bio-Gel A, Gelarose, Sagavac), which indicates the usefulness of this gel. It can be made with a very large pore size, and it is mechanically much more stable than a dextran gel of similar pore size.

Porous silica. Silica is commercially available in bead form (Porasil; Merck-O-Gel Si) with a range of porosities. This is an aerogel, with a very rigid structure. It can be used in some organic solvents, but it is best used in water. It is rather highly polar, and can tend to retard polar molecules by adsorption.

Porous glass. Various grades of granular porous glass (Corning; Bio-Glas) are available. Glass can be used for gel chromatography in both aqueous and organic media, but there may be undesirable adsorption effects, as with silica.

The gels formed from natural or synthetic organic materials are all fairly stable chemically, although the polysaccharide gels are hydrolysed in very acid solution, and some of the other polymers are unstable in very alkaline conditions. Most resins should be used only in the range pH 2 to pH 11. Polysaccharide resins are susceptible to microbial attack, and they should be sterilised for storage. Once the dry

Table 2.7 Some commercially available media for gel chromatography

Name	Type	Chemical nature	Eluant (mobile phase)	Maximum exclusion limit M_r	Calibration	Number of fractionation ranges	Bead size	Notes
Sephadex G	Xerogel	Dextran	Aqueous	6×10^5 2×10^5	Peptides/proteins Dextrans	12 8	10–40,* 40–120, 50–150, 100–200	Superfine grade for very high resolution and thin layers
Sephadex LH	Xerogel	Dextran hydroxypropyl ether	Polar organic	5×10^3		1	25–100*	Exclusion limit varies according to solvent
Bio-Gel P	Xerogel	Polyamide	Aqueous	4×10^3	Peptides/proteins	10	50–100,† 100–200, 200–400	
Enzacryl Gel K	Xerogel	Polyacryloyl-morpholine	Aqueous Polar organic	1×10^5	Polyethylene-glycols	2		Little swelling in lower alcohols
Bio-Beads SX	Xerogel	Polystyrene	Organic	14×10^3	Polystyrenes	7		
Styragel	Hybrid	Macroporous polystyrene	Non-polar organic	4×10^8	Polystyrenes	12		
Bio-Gel A	Hybrid	Agarose	Aqueous	15×10^7	Dextrans	6	50–100,† 100–200, 200–400	1, 2, 4, 6, 8, 10% agarose
Sepharose	Hybrid	Agarose	Aqueous	4×10^7 2×10^7	Proteins Polysaccharides	3	60–250,* 40–190, 40–210	2, 4, 6% agarose
Merck-O-Gel	Hybrid	Polyvinyl acetate	Polar organic	1×10^6	Polystyrenes	6		
Porasil	Acrogel	Silica	Aqueous Organic	2×10^6	Polystyrenes	6		Calibration in organic solvents
Bio-Glas	Aerogel	Glass	Aqueous Organic	9×10^6	Polystyrenes	5		Calibration in toluene

* μm
† U.S. Standard Screens.
M_r = relative molecular mass

beads have swollen the gel is best kept in that condition (this is essential in the case of agarose gels, which cannot be dried out). Packed columns can be stored and used repeatedly.

Gel media are made in a variety of pore sizes to give a series of fractionation ranges, and they are also made in several different bead sizes. Even when swollen the beads remain discrete from each other. The volume of solvent outside the beads, V_o, is available for the passage of the excluded molecules to travel through the column, and the rate of movement of those molecules will depend on the bead size, larger beads increasing the rate. Selection of an appropriate bead size depends on considerations rather similar to those described for ion-exchange chromatography (p. 49). Gels of very fine bead size are required for making gel thin-layer plates (p. 285).

Apparatus

Apparatus for gel chromatography is similar to that used for other forms of liquid chromatography on columns, but some manufacturers have designed apparatus specifically for use in gel chromatography, and in view of the unusual physical properties of gels there are advantages in using this special equipment.

The size of column which should be used depends on the volume of sample to be separated. For many purposes a bed volume of about five times the sample volume should suffice, with a ratio of diameter to height of about 1 : 15, although for very intricate separations the ratio may need to be as much as 1 : 100. Very long narrow columns are to be avoided, however, as in the usual downward flow long columns may easily become clogged. This can be avoided by using several shorter columns in series (Fig. 2.7), by shortening the column and recycling the eluate[69], or by using upward flow. Downward flow can be effected by gravity or by means of a pump; for recycling and upward flow the use

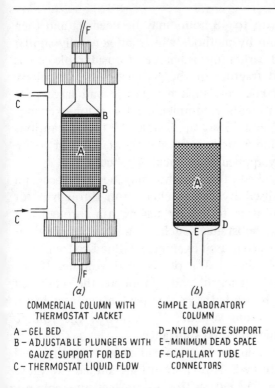

(a)
COMMERCIAL COLUMN WITH
THERMOSTAT JACKET

(b)
SIMPLE LABORATORY
COLUMN

A – GEL BED
B – ADJUSTABLE PLUNGERS WITH
GAUZE SUPPORT FOR BED
C – THERMOSTAT LIQUID FLOW

D – NYLON GAUZE SUPPORT
E – MINIMUM DEAD SPACE
F – CAPILLARY TUBE
CONNECTORS

Fig. 2.13. Columns for gel chromatography

of a pump (usually peristaltic) is essential. Large columns can be obtained for use on an industrial scale. One example is a multiple column made up from sections each having a bed volume of 16 dm³; single columns giving a bed volume up to 160 dm³ are made. A large unit of 2500 dm³ bed volume can desalt 1500 dm³ of solution per hour.

The column

Unless it be of the aerogel type, the stationary phase swells to a greater or less extent when it comes into contact with the solvent; the greatest swelling is with hydrophilic gels in water. The dry gel is therefore left to stand for a time in contact with the solvent (up to 48 hours may be needed) and then packed in a column by methods which are generally similar to those described earlier in this chapter. Considerable care is necessary to avoid fracture of the gel particles, particularly with the more fragile gels such as agarose and large pore xerogels. The treatments recommended for anion-exchange resins on pages 52 and 53 are applicable. All the precautions for packing a column with solid adsorbent referred to on page 20 are equally applicable to packing gel columns.

Having been packed, the column may be stabilised and tested. It is stabilised by having eluant run through it for several hours, and it is tested by having run through it a coloured excluded substance, such as a dyed protein or polysaccharide (an often-used example is 'Blue Dextran') in a sample volume of about 10% of the bed volume. If the column is properly packed and stabilised the coloured substance should be eluted as a compact horizontal band. This test gives the void volume, V_0, which is the volume required to elute the coloured substance.

The top of the column, as usual, must not run dry, and must not be disturbed. Any precautions similar to those described on page 21 will suffice to protect the column, although some authors recommend that the top of the column should not be covered by protective devices at all. With the softest gels it has been suggested that a shallow top layer of a low-porosity gel should be added[74]. In commercial apparatus where the column is contained by plungers no extra protection is needed.

The sample

Preparation. Solid particles must be removed from the sample, and also any substances which may be strongly adsorbed on the gel.

Volume. The mass and concentration of the solutes are not important in gel chromatography, except insofar as they affect the viscosity. The last is important, and should not be more than about twice that of the eluant. The sample volume is also important. The largest volume which can be handled for complete separation in one run through a column is $V_{e_1} - V_{e_2}$ where these are the elution volumes of the two most closely related components (from the chromatographic point of view). For group separations where $V_{e_1} - V_{e_2}$ is large, the sample volume can be 10–30% of the bed volume, V_t. For separation of closely related components ($V_{e_1} - V_{e_2}$ small) the sample volume should only be 1–3% of V_t. The smaller the sample volume, the greater will be the reduction of the component concentration in the eluate; this dilution effect may have to be taken into account in deciding upon column and sample sizes.

For very small samples, in the cubic millimetre region, gel chromatography may well be more effective on gel thin-layer plates than on columns (Chapter 5, p. 285). The thin-layer technique can often be used in a 'pilot' experiment to discover the best conditions for a larger-scale separation on a column.

Application. In simple laboratory columns the sample is applied at the top of the column by any of the methods already described in this chapter. A method not previously mentioned and suitable for application of protein samples such as blood serum is to leave a layer of the eluant above the top of the column and to introduce the sample through a capillary on to the top of the column below the eluant surface. For this to work the sample solution must be denser than the eluant solution.

Commercial plunger-type columns have special inlet arrangements for the sample.

Applications

Since 1959 many hundreds of papers have been published on the use of gel chromatography, and it is therefore only possible to summarise a few applications here. Some typical separations are listed in Table 2.8. Although the method is mainly used for small-scale separations for research and routine analysis, there are larger-scale uses in industrial production as well.

Gel chromatography was first used to separate large molecules of biological origin, such as proteins, poly-saccharides, nucleic acids, enzymes, and it still finds its most widespread application in those areas. More recently the separation and examination of synthetic polymers has been added to the field of application, and the method has become an important part of polymer technology.

Desalting of solutions, such as, for instance, of proteins, is an important use of gel media. Fractionation of large molecules, as for example, of serum proteins, is carried out on an analytical and a preparative scale, and when the molecules concerned are chemically related the fractionation curve can be used as the basis for determining M_r and M_r ranges. This has been done, for instance, for proteins[82, 83], polysaccharides, and synthetic polymers.

The range of relative molecular mass over which separations have been achieved is very large. Two extreme examples are given in Table 2.8. The separation of oligoethylene glycols and the monomer is an example of the fractionation of relatively small molecules, and the separation of polio virus and influenza virus an example of a separation of species so large as to be regarded as particles rather than molecules.

Industrial use of gel chromatography, other than as an analytical tool in research and process control, is largely confined to desalting; the pharmaceutical industry, and, to a lesser extent, the food industry, are the main users.

Gel ion-exchangers

Many large molecules of biological origin are polar in nature and hence it should in principle be possible to separate them

Table 2.8 *Some examples of the use of gel chromatography*

Gel medium	Mobile phase	Separation	Notes	Ref.
Sephadex G-10	Phosphate buffer pH 7.0	Ethylene glycol polymers and oligomers	M_r range 62–600; flowrate 6 cm³ hr⁻¹	75
Sephadex LH-20	Chloroform	Triglycerides		76
Sephadex G-200	pH 8.0 buffer	Human serum proteins		77
Sephadex G-75 G-100 G-200	0.05 mol dm⁻³ sodium acetate buffer	Enzymes of dog pancreas: lipase, ribonuclease, trypsinogen, amylase	G-75 gave the best separation	78
Agar beads (2.5% agarose)	Salt solution (NaCl, KCl, KH_2PO_4, $CaCl_2$, $MgCl_2$)	Polio virus and influenza virus	Virus sizes 28 μm and 100 μm	79
Bio-Beads S-X1 + S-X2	Benzene	Triglycerides and hydrocarbons	Two columns in series	80
Sephadex LH-20	Methanol	Cyclic caprolactam oligomers	Two columns in series	80
Bio-Gel A-50m	pH 8.9 buffer	Pig serum lipoproteins	Flowrate 50 cm³ hr⁻¹	81
Bio-Gel P-100 P-200 P-300	pH 7.4 and pH 6.8 buffers	M_r of proteins: insulin, cytochrome C, ovalbumin, serum albumin	Uncertainty in values approx. 12%	82
Sephadex G-200	pH 7.5 buffer	M_r of proteins: 31 proteins examined	M_r range: glucagon (M_r 3.5×10^3) to bovine α-crystallin (M_r c. 800×10^3)	83
Aquapak	Water	Dextrans	Flowrate 1.1 cm³ min⁻¹ at 44°C	84
Styragel	Tetrahydrofuran	Normal hydrocarbons		84
Bio-Glas	Perchloroethylene	Polyethylenes		85
Silica	Tetrahydrofuran	Polystyrenes, polyvinylchlorides	Separation at 110°C	86
	Water	Dextrans		
Sephadex LH-20	Tetrahydrofuran	Tar/bitumen mixtures		114

by ion-exchange chromatography. It is, however, not normally practical to separate such molecules on conventional ion-exchange materials because the molecules cannot penetrate the resin matrix, even if cross-linking is as little as 1%. Most of the exchange sites are therefore inaccessible and, since the molecules do not use the full exchange capacity of the material, the separation is very inefficient. Introduction of ion-exchange groups into gel media[87] has overcome this difficulty, and has made possible the ion-exchange separation of large molecules such as proteins.

Since a material with a large pore size is needed, it is not surprising that introduction of ionising groups into dextran polymers has given the best results. The only readily available ion-exchange gel media are dextran ethers, in which the −OH groups have been converted into −OR, where R contains either an acidic or a basic group. Some of the properties of ion-exchange dextrans are listed in Table 2.9.

Table 2.9 *Commercially available dextran ion-exchangers*

Name	Type	Functional group	Exchange capacity mmol H^+ (OH^-) g^{-1} (approx.)
SP-Sephadex	Strong acid	Sulphoxyl	2.3
CM-Sephadex	Weak acid	Carboxymethyl	4.5
QAE-Sephadex	Strong base	Diethyl-(2-hydroxy propyl)aminoethyl	3.0
DEAE-Sephadex	Weak base	Diethylaminoethyl	3.5

Physical properties. Ion-exchange dextrans are physically very similar to ordinary dextrans, including their chemical stability, but the extent of swelling in aqueous buffers is very much more dependent on the ionic strength. When the ionic strength of the solution is lowered the gels swell considerably; they shrink again when the ionic strength is raised. This is of practical importance in the operation of a column since if a considerable shrinkage occurs when a buffer is changed cavities and irregularities may appear in the bed. Again, excessive swelling may clog the column. Swelling is most likely to happen during regeneration, and in consequence it is often better to regenerate the resin in a Büchner funnel, and then to repack the column.

Selection of a suitable gel. In many cases it is clear whether an anion- or a cation-exchanger is required for a separation, but some important macromolecules of biological origin are amphoteric electrolytes (such as proteins). At a pH above the isoelectric point they are in the anionic form, and are separated on an anion-exchanger; below the isoelectric point they are cationic, and a cation-exchanger is used. The pH of the buffer is chosen according to the nature of the substances to be separated, and this then dictates the gel type to be used. If, at the pH selected, the ampholyte has a high affinity for the exchange group of the resin the appropriate weak ion-exchanger should be used; and if the affinity is low a strong ion-exchanger should be used.

Ion-exchange dextrans are supplied in two degrees of porosity, the less porous being recommended for separations where M_r is less than 3×10^4 or greater than 2×10^5 (in the latter case the molecules will be totally excluded from the gel and little ion-exchange will take place). The more porous gel is recommended for separations between the two M_r limits stated. The bead size in all cases is $40-120$ μm.

Technique. Ion-exchange gels are used and handled in much the same way as are the unmodified gels, combined with some of the techniques used for conventional ion-exchange materials. Allowance must be made for changes in the gel condition when the ionic strength of the eluant is changed. Ion-exchange gels are supplied in the sodium form (cation-exchangers) or the chloride form (anion-exchangers) and they are usually converted, before a separation is started, into the H_3O^+ or OH^- form respectively.

Applications. The number of reported uses of ion-exchange dextrans is increasing rapidly. Examples are the use of DEAE-Sephadex for the separation of mucopolysaccharides by stepwise elution with NaCl in HCl[88], soluble RNA[89], and the separation of myoglobin and haemoglobin[90]. Cation-exchange dextrans have been used for the separation of β- and γ-globulin[91], alkaloids[91], and for the purification of enzymes[92].

Affinity chromatography

Ordinary chromatographic techniques depend for their effectiveness on differences (often small) in adsorption, partition, ionic charge, or size, between solute molecules of similar chemical character. A modified gel technique in which use is made of the high specificity of biochemical reactions is affinity chromatography[93, 94].

An example of the use of affinity chromatography might be the extraction of an enzyme inhibitor from a biological fluid. The principle is illustrated diagrammatically in Fig. 2.14. A ligand, in this case the enzyme, is covalently bound to the gel medium by a suitable chemical reaction. A column is then packed with this modified gel, and the sample is introduced in the usual way, with a suitable buffer in which the substrate, in this case the inhibitor, is specifically adsorbed by the ligand. Elution with that buffer then removes the unwanted substances from the column, and a change in the eluant composition reverses the enzyme-inhibitor binding and the inhibitor can be eluted. The conditions which must be fulfilled are therefore: the gel-ligand bond must be stable under the conditions of the experiment; the ligand-substrate bond must be specific, and reversible under the conditions of the experiment; and the gel-ligand bonding must be such that the specific adsorption properties of the ligand are not interfered with by the gel matrix.

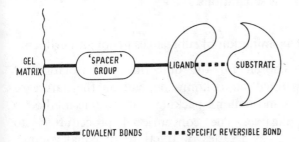

Fig. 2.14. Principle of affinity chromatography

It has been found that small ligands retain their specificity to a much greater extent if a 'spacer' group, such as a chain of about six carbon atoms, is incorporated between the gel and the ligand[95] (Fig. 2.14).

Most of the work so far reported has been with polysaccharide gels (usually agarose), although there are instances of the use of polyamides. The usual way of coupling a ligand to a polysaccharide gel is to treat the gel with cyanogen bromide in aqueous alkali, to give a reactive intermediate which can be isolated and then made to react with a variety of ligand molecules[96]. The intermediate can also be made to react first with a simple molecule to introduce a spacer group, and the ligand is then coupled in a subsequent step.

These reactions may require some skill in organic synthesis, and also considerable expenditure of time. Several intermediates are available commercially; for instance, CNBr-Sepharose, the product of reaction between Sepharose and cyanogen bromide; Sepharose and Bio-Gel P (polyamide) intermediates with reactive spacer groups attached; and Sepharose already coupled to an adsorbent specific for polysaccharides and glycoproteins.

The types of system for which affinity chromatography is used are bochemical and clinical; for example, enzyme-inhibitor, antibody-antigen, hormone-carrier, and also processes such as virus separations.

Detection and examination of substances in column effluents

The methods chosen for the detection and isolation of substances separated on columns depend on the nature of both substances and column packing, and no one method is universally applicable. The conventional procedure is to collect the eluate as a series of fractions of equal volume, usually with an automatic fraction collector, and then to examine each fraction by a method such as spectrophotometry, refractive index or conductivity measurement, polarography, or classical chemical methods. Continuous monitoring of the eluate can be arranged, although it requires more sophisticated equipment, both in the chromatographic columns and in the means of detection.

Collection of fractions may be avoided in simple cases where the components of the mixture have separated on the column when the fastest-moving one has reached the end. If the elution is stopped at that point the substances may be detected on the column — if they are not visibly coloured, by observation in ultraviolet light. The column packing may then be pushed out and the sections containing each substance may be separated and eluted individually. Alternatively, after the column has been extruded it may be treated with a locating reagent in the manner of paper and thin-layer chromatography (pp. 149 and 301) to locate the substances. If flexible plastics tubing is used, and the substances can be detected on the column, extrusion is neither necessary nor practicable; the whole tube is simply cut into appropriate sections with a sharp knife.

In dry column chromatography extrusion or cutting-up are almost essential. In wet column methods it is not usual to extrude any but adsorption columns, since it is essential that the extruded column shall retain its shape when taken from its container. Partition, ion-exchange, and gel columns can be effectively used only with the fraction-collecting or continuous monitoring systems. If actual specimens of the separated substances are required separate fractions must be collected. Continuous monitoring, with automatic recording of effluent composition, is an integral part of the system of high-efficiency liquid chromatography described in the ensuing section of this chapter.

High performance liquid chromatography

Possibly the most significant advance in chromatography since 1966 is the development of liquid chromatography along lines similar to those followed by gas chromatography. The technique now differs so greatly from the older methods that it is referred to as high performance or high efficiency liquid chromatography. Significantly, gas chromatography, although introduced more recently than liquid chromatography, has always had a firm theoretical basis, the rate theory of van Deemter, Klinkenberg and Zuiderweg[97] having been especially useful. Liquid chromatography on the other hand, in spite of early spectacular successes, stagnated somewhat through lack of an adequate theory on which to base improvements in column performance and, to a lesser extent, also on account of the limited detection systems available until recently.

Current practice in high performance liquid chromatography is based on the application to liquid systems of the concepts inherent in the van Deemter theory (above) and embodied in the so-called van Deemter equation. A more detailed account of this equation appears in Chapter 4 (pp. 268 – 272) but the equation in its simplified form is reproduced here for convenience:

$$\text{Height Equivalent to a Theoretical Plate} = \text{HETP} = A + \frac{B}{\bar{u}} + C\bar{u}$$

where $\bar{u}$ is the average linear velocity of the solvent through the column (for liquids u, the linear velocity, is substantially constant throughout the column length, in contrast to gas chromatography where u varies continuously).

The detailed application of this equation is more complicated for liquid systems than for gaseous ones hence the following is a very much simplified account, but it is hoped that it will form an adequate introduction to modern liquid chromatography practice. Very useful reviews are given by Huber[98], Perry, Amos and Brewer[99], Perry[100], Snyder[101] and Kirkland[102], and the literature is comprehensively documented in Gas and Liquid Chromatography Abstracts[103].

The length of the column l divided by the HETP (see p. 264) gives the plate number n, and the greater the value of n the greater the separating power of the column. The most obvious way, therefore, of improving the performance of a column is simply to lengthen it. This accounts in part for the adoption of long thin columns in high efficiency systems, as compared with the short, relatively large bore columns described earlier in this chapter (p. 19). In order to make maximum use of a given length of column, however, the HETP needs to be as small as possible and this means that the values of the constants A, B, and C in the van Deemter equation need to be minimised and the value of $\bar{u}$ optimised. A plot of the van Deemter equation – HETP against $\bar{u}$ – is a hyperbola, and this has been found to be true using experimental values in gas chromatography – as in the upper curve in Fig. 2.15, and on p. 269 – but the corresponding experimental curve for liquid chromatography is slightly different in shape, as shown in the lower curve in Fig. 2.15. Also, in liquid systems, it has only been found practicable to work over a range of solvent velocities – and hence flow-rates – as indicated approximately by the solid line of the lower curve in the figure:

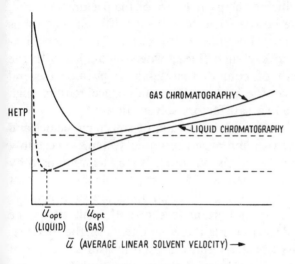

Fig. 2.15. *Plot of van Deemter equation (curves separated for clarity)*

Furthermore, the lowest flowrates investigated are impracticably slow although still giving solvent velocities higher than $\bar{u}_{opt}$, hence a solvent velocity greater than the optimum is invariably employed. In order to minimise this disadvantage it is important to ensure that the slope of the solid line in the lower curve in Fig. 2.15 is as small as possible so that a given practical solvent velocity is as close as possible to $\bar{u}_{opt}$. The main factor determining this slope is C in the van Deemter equation. Ways in which the constants A, B, and C may be minimised will be discussed.

Reduction of HETP. The column will be working most

efficiently when the HETP is a minimum, and when A, B, and C in the van Deemter equation are a minimum.

The constant A arises from the different lengths travelled by the solute molecules and depends on the shape and size of the column packing. It is probably the main contributor to band spreading and is a constant for a given column. It can be reduced by reducing the particle size of the packing material, hence very fine powders are used in high efficiency columns (10–20 μm, preferably spherical in shape for regular packing). Needless to say such fine powders – coupled with the increased length of column that tends to be used (several metres) – reduce the permeability and hence require high pressures to produce satisfactory solvent flows.

B accounts for the axial diffusion of solute molecules and is probably not very important in liquid systems – certainly not as important as in gaseous ones. It may be minimised by making the mobile phase as viscous as possible, but this must be paid for in terms of the higher pumping pressures needed. It is, perhaps, more important to choose the solvent on the basis of selectivity towards the components of the mixture.

C is the mass transfer term, that is, it measures the rate of transfer of solute molecules between the mobile and the stationary phases. For this rate to be as high as possible, and hence C as small as possible, it follows that the particle size of the column packing material needs to be small (the same requirement as for minimising A), and if the particle is coated with a layer of liquid as in partition chromatography, the thickness of the layer should be made as small as possible, consistent with an acceptable capacity. This concept, along with certain other requirements of packing materials in high pressure systems has led to a new generation of stationary phases named pellicular materials, where a thin porous layer is held on a solid, impermeable core. These materials are described in more detail later. The mass transfer term is significantly larger with gel columns than with adsorption or partition systems.

In brief, a high efficiency liquid chromatograph is characterised by a long, narrow column packed with very fine powder with spherical particles, operated at high liquid pressures, and possessing a continuous monitoring facility for the column effluent. Such chromatographs may be quite expensive, but in return it is possible to achieve efficiencies comparable with those obtainable in gas chromatography, with separating times of the same order, and the ability to work with non-volatile materials.

Apparatus and technique

Technically, development work has concentrated on the production of pulse-free pumps for generating the high pressures required, and sensitive detectors for monitoring the column effluent. In addition considerable work has gone into the production of column packing materials that can be used at high pressures – as already mentioned.

The use of high pressures is not the hazard that it may seem because the pressures are hydrostatic, and should any part of the system fracture under stress, pressure is immediately relieved throughout the system. The fact that the pressure is hydrostatic is an advantage in another way; it means that there is very little pressure drop across the column and hence the linear velocity of the mobile phase through the column is virtually constant from point to point. Thus if the optimum – or as near optimum as can be achieved – linear velocity of mobile phase through the column is employed for maximum efficiency, this velocity remains constant throughout the column length, unlike gas chromatographic systems where only a relatively small portion of the column is utilised at maximum efficiency owing to the compressibility of the gas phase – see p. 270.

Fig. 2.16 shows schematically an arrangement suitable for high pressure liquid chromatography, and it is interesting to compare this arrangement with that in Fig. 4.1 (p. 182) for

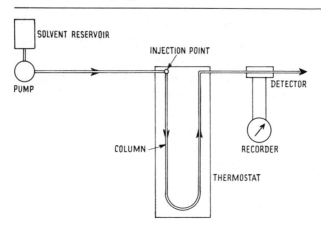

Fig. 2.16. Apparatus for liquid chromatography schematic

gas chromatography the main difference being – apart from the difference between the mobile phases – the inclusion of a pump in the liquid system. In designing and assembling the above apparatus very careful attention is paid to minimising 'dead' volumes, that is the internal volume of the apparatus (apart from the solvent reservoir) not occupied by column packing. Accounts of the equipment have been given by Conlon[104] and Zlatkis[105].

A brief discussion of the various parts of a liquid chromatograph follows.

Pumps. An important requirement of a pump for liquid chromatography is that it should be pulse-free since detectors are usually flow-sensitive (although pulses can be smoothed out to a large extent by the inclusion of one or more pressure gauges in the line). It should also produce pressures up to about 3.5×10^6 kg m^{-2} (5000 psi), and be compatible with solvents, continuous in operation and of small volume – not greater than, say, 0.5 cm^3. Satisfactory pumps are available commercially.

Column. Columns and lines generally are probably best made in stainless steel, although thick-walled glass columns are manageable. The advantages of stainless steel are largely those manifest in gas chromatography, namely, that it is robust and that columns made from it can be bent to a convenient shape to fit a thermostat. Columns vary in diameter from 1 to 4 mm and the length may be up to 50–60 m. If possible it is best to pack columns in a straight length, or lengths, and bend them to a convenient shape afterwards.

Packing the column. Guidance on packing the column is disappointingly sparse in the literature, particularly since this is a crucial operation and the fine powders involved make it a difficult operation. A dry-column technique is recommended and Perry *et al*[99] describe a method in which dry packing, sufficient to fill about 2–3 cm of column is added and tamped into place using a teflon-tipped plunger. After consolidation a further addition of packing material is made, and the process repeated until the column is full. Columns up to 1 m are packed in this way, and if a greater length is needed further columns are packed by the same procedure and coupled in series until the desired length is achieved.

Detectors. As mentioned at the beginning of this chapter, the detection of substances in liquid column effluents is not as straightforward as in gas chromatography. The advantages of high efficiency liquid chromatography have, however, stimulated the development of suitable detectors and with considerable success, although some workers had already made progress along these lines for monitoring effluents from columns operating under normal pressure conditions.

The earliest, and most tedious, technique was to use a fraction collector and to analyse the contents of each fraction separately by chemical or physical means. Claesson[106] describes a method for the continuous monitoring of refractive index. Spackman, Stein and Moore[107] developed a method (now in use in certain commercial instruments) for amino-acids in which the column effluent is met by a stream of ninhydrin reagent. The characteristic ninhydrin colour is developed by passing the mixture through a heated capillary, and the absorbance of the resulting solution is measured continuously. Blaedel and Todd[108] used a polarographic method for the continuous analysis of reducible substances, and Johansson, Karrman and Norman[109] developed a high frequency technique for the continuous analysis of bile acids separated on a reversed phase partition column. More recently Lovett and Wright[110] described a method for continuously measuring the transmission of ultraviolet light by column effluents containing nucleotides and similar compounds.

The above work to some extent anticipated recent developments, a review of which is given by Conlon[104].

Ideally a detector should be sensitive (of the order of $0.1 \ \mu g \ cm^{-3}$), show a linear response with concentration over a wide range for all substances, possess a small dead volume, be non-destructive, and be continuous in operation. Additionally, the ideal detector would be insensitive to ambient conditions. Needless to say no instrument fills all these requirements, and most are expensive into the bargain.

The most popular types of detector used in commercial instruments to date include those employing: ultraviolet absorption; measurement of refractive index; heat of adsorption (or solution); the moving wire system and electrolytic conductivity. Other properties that are being examined include: dielectric constant, fluorescence, and polarography. *Ultraviolet and visible absorption.* The ultraviolet detector can be very sensitive but has obvious limitations to certain classes of compounds, namely those that exhibit extinction in some region of the ultraviolet spectrum or, for visible absorption, give a colour with a suitable reagent (cf Spackman, Stein, and Moore[107]). One particular advantage, however, is that it can sometimes determine unresolved compounds.

Clearly the solvent should not absorb at the chosen wavelength, and hence gradient elution may be used with suitable solvents without affecting the detector.

Refractive index. The sensitivity of refractive index detectors is high and the system is not limited to particular classes of compounds. The sensitivity is, however, limited somewhat if gradient elution is used. To compensate for the changing refractive index in gradient elution a differential system may be employed where two effluent streams with identical solvent composition are compared, but only one − from the column − carries the separated components of the mixture.

Unlike the ultraviolet detector, the refractive index detector is very temperature sensitive, and on this account is also very sensitive to flowrate of column effluent since changes can upset the thermal equilibrium.

One of two techniques may be used, one measuring the bending of a beam of monochromatic light, and the other measuring the intensity of the light reflected from the surface of the column effluent where the intensity varies inversely as the refractive index with constant incident light. The former method requires more expensive instrumentation than the latter.

Heat of adsorption. Detectors employing heat of adsorption are of fairly general application and good sensitivity. Small temperature changes arising from the interaction between the components of the sample and the column packing are detected by, for example, a thermistor embedded in the packing material near the end of the column. As each component passes the sensor, heat is first evolved, and then absorbed, corresponding to adsorption and then desorption. As a result the recorded peak shape is first positive and then negative − see Fig. 2.17.

In practice the detector works equally well with adsorption − including ion-exchange − partition and gel columns. The peak height, positive or negative, is proportional to the

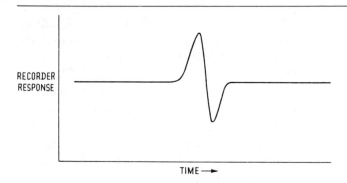

RECORDER
RESPONSE

TIME →

Fig. 2.17. Record from a heat of adsorption detector

concentration of the component, but frequent calibration appears to be necessary owing to changes that take place in the stationary phase from run to run. It follows from the mode of operation of the detector that it should be maintained at as constant a temperature as possible and that the flowrate must be constant. Gradient elution may be used but causes base-line drift to a greater or lesser degree.

Moving wire detection system. The argon and flame ionisation detectors have been employed for many years in gas chromatography (p. 199) and use has been found for their high sensitivty in the monitoring of liquid column effluents. The system was devised by Scott and James[111] and consists of a moving stainless steel wire which passes through a cleansing oven in which surface impurities are removed by means of a stream of argon. The wire next passes through the column effluent and is coated with a film of the solution emerging. Next the wire passes through another oven maintained at an appropriate temperature, that is, sufficient to remove solvent (mobile phase) but not solutes. Finally the wire passes into a pyrolysis chamber where the solutes are thermally cracked and swept into the ionisation detector − in the case of the argon detector with a stream of argon. The device is shown schematically in Fig. 2.18.

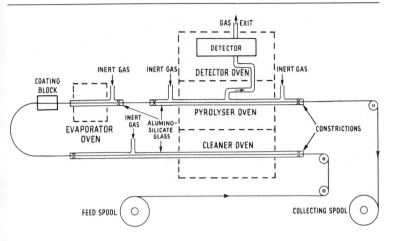

Fig. 2.18. Moving wire detection system

Limitations of the system are that it will not respond to many inorganic substances (see p. 200), and others that do not give rise to pyrolysis products capable of ionisation in the detector. The system also cannot cope with substances of volatility comparable to that of the mobile phase. A robust instrument based on the above has been commercially developed.

The flame ionisation detector has the advantage over the argon detector that it may be operated at a higher temperature thus avoiding condensation of some pyrolysis products. *Conductivity.* The specific conductivity of the column effluent is monitored by means of platinum electrodes contained in a low-volume cell through which the eluate passes immediately after leaving the column. An alternating voltage (60—1000 Hertz at 14 V) is applied and the cell forms one arm of a Wheatstone bridge, the voltage necessary to restore the balance being measured. Both aqueous and non-aqueous media can be handled.

Principal advantages of the detector are simplicity, robustness and a linear response over a wide range of concentration, hence its usefulness in quantitative work. Constant temperature and constant flowrate are desirable, and gradient elution will upset the baseline and is best avoided where quantitative work is important.

Identification of separated components

The detectors mentioned above respond to the presence of a component in the effluent stream and their response is fed to a chart recorder which produces a chromatogram. An example is shown in Fig. 2.19.

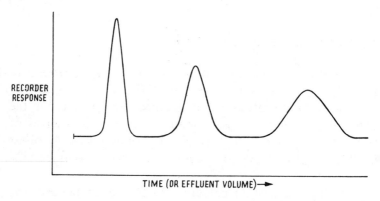

RECORDER
RESPONSE

TIME (OR EFFLUENT VOLUME)——➤

Fig. 2.19. Chromatogram produced by differential detector

Such a record is produced by a so-called 'differential' detector – cf p. 192 – although as already noted the heat of sorption detector produces peaks that are at first positive and then negative. To identify the components giving rise to the various peaks the column effluent fractions containing them may be collected and subjected to the usual identification procedures – if the quantities allow – such as infrared and ultraviolet spectroscopy, nuclear magnetic resonance spectrometry and so on. Alternatively the retention volume V_R may be used in a similar manner to that described on p. 206

in gas chromatography, with similar restrictions on their applicability to symmetrical peaks only.

Estimation of separated components

Quantitative measurements are readily carried out on the chromatogram since the amount of substance present is proportional to its peak area. Methods of measuring the area are given on p. 216. Calibration is necessary, that is, peak areas need to be correlated with known sample sizes, bearing in mind the fact that the detector response may well be different for different molecular types. For this reason the analysis of standard samples is often necessary for comparative purposes. As with gas chromatography an on-line computer system may be used with great advantage, if available. Such an arrangement may be made not only to detect peaks but to determine their areas and at the same time correct for variations such as base-line drift and incomplete resolution.

New stationary phases

It is generally assumed (though some workers have disputed this) that uniform bead materials are particularly good for packing columns because they lead to homogeneous packing and good permeability. This has led to the production of such materials in the form of porous beads for chromatography – both in the liquid phase and the gas phase. Again, many materials satisfactory for conventional liquid chromatography are of little value in high speed, high efficiency columns because of their inability to withstand the pressures employed without serious deformation or even collapse. Additionally, the column packing should be stable to changes in liquid concentration and temperature, that is, it should not swell or shrink.

Pellicular materials, also known as porous-layer beads or controlled surface porosity supports (that is, solid impervious beads of high mechanical strength made from a material such as stainless steel, glass, or a highly cross-linked polymer, surrounded by uniform porous shells (Fig. 2.20)) are particularly effective, and are commercially available. They differ from ordinary porous beads in that their capacity is very much smaller, confined as it is to the surface layer, hence column loadings must be smaller, but they are capable of higher efficiencies because the rate of mass transfer between the mobile phase and the porous layer is greater (small C value in the van Deemter equation).

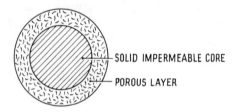

SOLID IMPERMEABLE CORE

POROUS LAYER

Fig. 2.20. Bead of pellicular material

Horvath and Lipsky[112] have described methods for the production of such materials and these include:

(*i*) The solid impermeable beads are kneaded with a suspension of fine particles with simultaneous evaporations of the solvent. The shell of deposited material may be sintered in situ; for example, Baymal, a colloidal gamma-alumina (see p. 246) can be treated in this way by heating to $450°C$.

(*ii*) Deposition by chemical reaction; for example, beads coated with a suspension of cellulose derivatives such as cuproxam or viscose can be suspended in a solvent and the cellulose regenerated on the surface. Similarly, cross-linked polystyrene can be deposited and subsequently converted to ion-exchange resin.

(*iii*) Precipitation of solids on to beads. Silicic acid may be precipitated as a porous layer by the acid hydrolysis of a silicic acid ester. Subsequent heating stabilises the shell and increases the activity, provided that the temperature is not too high.

Chemically bonded stationary phases

Packing materials, including ordinary porous beads as well as pellicular materials, may be used on their own as stationary phases in liquid-solid chromatography (ion-exchange materials may be included here), or they may be impregnated with a liquid stationary phase to give a partition system. There is, however, a danger of the liquid phase being stripped by the high shearing action of the mobile phase, which is under high pressure, if high flowrates are used, and by the normal solvent processes. The latter effect may be minimised by pre-saturating the mobile phase with the stationary phase by passing it through a pre-column, but the presence of the stationary phase in the mobile phase is sometimes undesirable because of possible interference with the efficiency of the detector. Additionally, the pre-saturation technique is less satisfactory if gradient elution is to be used. Chemically bonded stationary phases have therefore been developed and they overcome most of the difficulties just mentioned. Such stationary phases are chemically bonded to the surface of the packing material and, with careful choice of mobile phase and the use of reasonably low temperature – not more than about $50°C$ – are not readily stripped.

Generally, the surface bonded materials are in the form of silicate esters or silicones. Thus alcohols will react with surface hydroxyls of silica to give, for example, oxydipropionitrile and n-octyl esters. Organochlorosilanes also react readily with surface hydroxyls (see also p. 248) to give, for example, octadecylsilane groups bonded to the surface. A useful review of the above packing materials has been given by Majors[113].

Comparison of gas chromatography with high performance liquid chromatography

The principal disadvantages of using a liquid mobile phase as compared with a gaseous one are those of detection of separated components and the lower rate of mass transfer during the separation process.

Detectors for liquid chromatography are generally much more expensive and sometimes of more limited applicability than their gas chromatographic counterparts. The limits of detection do not, as yet, approach those of the most sensitive ionisation detectors used in gas chromatography.

Rates of diffusion in liquid systems are slower than those in gaseous ones, hence rates of mass transfer between the mobile and the stationary phases are slower, thus limiting the speed with which a column may be operated under anything approaching equilibrium conditions. Equilibrium between the mobile and the stationary phase is also complicated by the nature of the mobile phase which can usually interact quite strongly with the stationary phase, although this interaction can be turned to advantage in certain situations as in gradient elution.

Advantages of high performance liquid chromatography – apart from the fact that involatile materials can be handled – include the ability to operate long columns at near optimum solvent velocity throughout their whole length, and the possibility of varying the nature of the mobile phase to produce either stepwise or gradient elution, or both.

Supercritical liquid chromatography

A new technique involving the use of gases at very high pressures but at temperatures above their critical temperature has been developed. It may be regarded as intermediate between high performance liquid chromatography and gas chromatography, but is, perhaps, more closely related to the latter and hence a brief account is included in Chapter 4, p. 257.

REFERENCES

1. Kuhn, E., Winterstein, A., Lederer, E., *Z. physiol. Chem.*, 1931, **197**, 141.
2. Alm, R. S., Williams, R. J. P., Tiselius, A., *Acta Chem. Scand.*, 1952, **6**, 826.
3. Trappe, W., *Biochem. Z.*, 1940, **305**, 150.
4. Williams, R. J. P., Hagdahl, L., Tiselius, A., *Arkiv Kemi*, 1954, **7**, 1.
5. Strain, H. H., *Chromatographic Adsorption Analysis*, Interscience, New York, 1942.
6. Bickoff, E. M., *Analyt. Chem.*, 1948, **21**, 20.
7. Knight, H. S., Groennings, S., *ibid.*, 1954, **26**, 1549.
8. Seki, T., *Nature*, 1958, **181**, 768.
9. Brockmann, H., Schodder, H., *Ber.*, 1941, **74**, 73.
10. Hernandez, R., Hernandez, R., Jr., Axelrod, L. R., *Analyt. Chem.*, 1961, **33**, 370.
11. Sing, K. S. W., Madeley, J. W., *J. Appl. Chem.*, 1953, **3**, 549.
12. Madeley, J. W., Ph.D. Thesis, London, 1957.
13. Neish, A. C., *Canad. J. Chem.*, 1951, **29**, 552.
14. Moyle, V., Baldwin, E., Scarisbrick, R., *Biochem. J.*, 1948, **43**, 308.
15. Zbinovsky, V., *Analyt. Chem.*, 1955, **27**, 764.
16. Martin, A. J. P., Synge, R. L. M., *Biochem. J.*, 1941, **35**, 1358.
17. Moore, S., Stein, W. H., *J. Biol. Chem.*, 1949, **178**, 53.
18. Smith, G. H., *J. Chem. Soc.*, 1952, 1530.
19. Martin, A. J. P., Porter, R. R., *Biochem. J.*, 1951, **49**, 215.
20. Daly, M. M., Mirsky, A. E., *J. Biol. Chem.*, 1949, **179**, 981.

21. Edwards, R. W. H., Kellie, A. E., *Chem. and Ind.*, 1956, 250.
22. Hoffman, E., Johnson, D. F., *Analyt. Chem.*, 1954, **26**, 519. Johnson, D. F., Hoffman, E., Francis, D., *J. Chromatog.*, 1960, **4**, 446.
23. Hoffman, E., Johnson, D. F., Hayder, R., *Acta Endo.* (Copenhagen), 1956, **23**, 341.
24. Hathway, D. E., *J. Chem. Soc.*, 1958, 520.
25. Grassman, W., Deffreer, G., Schuster, E., Pauckner, W., *Chem. Ber.*, 1956, **89**, 2523.
26. Partridge, S. M., Swain, T., *Nature*, 1950, **166**, 272.
27. Burstall, F. H., Davies, G. R., Wells, R. A., *Discuss. Faraday Soc.*, 1949, **7**, 179.
28. Smith, A. I., *Analyt. Chem.*, 1959, **31**, 1621.
29. Sauer, R. W., Washall, T. A., Melpolder, F. W., *ibid.*, 1957, **29**, 1327.
30. Wager, H. G., Isherwood, F. A., *Analyst*, 1961, **86**, 260.
31. Siekierski, S., Fidelis, I., *J. Chromatog.*, 1960, **4**, 60.
32. Wren, J. J., *ibid.*, p. 173.
33. Alberti, G., Grassini, G., *ibid.*, p. 86.
34. *idem, ibid.*, p. 83.
35. *idem, ibid.*, p. 425.
36. Strelow, F. W. E., *Analyt. Chem.*, 1960, **32**, 1185.
37. *Dowex: Ion Exchange*, The Dow Chemical Company, Midland, Michigan, 1958, p. 9.
38. Spedding, F. H., *Discuss. Faraday Soc.*, 1949, **7**, 214.
39. Marcus, Y., Nelson, F., *J. Phys. Chem.*, 1959, **63**, 77.
40. Thompson, S. C., Harvey, B. G., Choppin, G. R., Seaborg, G. T., *J. Amer. Chem. Soc.*, 1954, **76**, 6229. Choppin, G. R., Harvey, B. G., Thompson, S. C., *J. Inorg. Nucl. Chem.*, 1956, **2**, 66.
41. Hulet, E. K., Gutmacher, R. G., Coops, M. S., *ibid.*, 1961, **17**, 350.
42. Samuelson, O., Sjoberg, B., *Analyt. Chim. Acta*, 1956, **14**, 121.
43. Wish, L., *Analyt. Chem.*, 1961, **33**, 53.
44. Hague, J. L., Machlan, L. A., *J. Res. Nat. Bur. Stand.*, 1961, **65A**, 75.
45. Bandi, W. R., Buyok, E. G., Lewis, L. L., Melnick, L. M., *Analyt. Chem.*, 1961, **33**, 1275.

46. Fritz, J. S., Garralda, B. B., Karraker, S. K., *ibid.*, p. 882.
47. Moore, S., Stein, W. H., *J. Biol. Chem.*, 1951, **192**, 663; 1954, **211**, 893.
48. Wade, H. E., *Biochem. J.*, 1960, **77**, 534.
49. Oliviero, V. T., *Analyt. Chem.*, 1961, **33**, 263.
50. Skelly, N. E., *ibid.*, p. 271.
51. Tomsett, S. L., *Analyt. Chim. Acta*, 1961, **24**, 438.
52. Peterson, E. A., Sober, H. A., *J. Amer. Chem. Soc.*, 1956, **78**, 751.
53. Rieman, W. J., *J. Chem. Ed.*, 1961, **38**, 339.
54. Amphlett, C. B., *Inorganic Ion-Exchange Materials*, Elsevier, Amsterdam, 1964.
55. Marshall, G. R., Nickless, G., *Chromatographic Revs.*, 1964, **6**, 180.
56. Thistlewhaite, W. P., *Analyst*, 1947, **72**, 531.
57. Smit, J. van R., Robb, W., Jacobs, J. J., *J. Inorg. Nucl. Chem.*, 1959, **12**, 104.
58. Smit, J. van R., *ibid.*, 1965, **27**, 227.
59. Smit, J. van R., Robb, W., *ibid.*, 1964, **26**, 509.
60. Bactslé, L., Paelsmaker, J., *ibid.*, 1961, **21**, 124.
61. Amphlett, C. B., MacDonald, L. A., Burgess, J. S., Maynard, J. C., *ibid.*, 1959, **10**, 69.
62. Coleman, C. F., Blake, C. A., Brown, K. B., *Talanta*, 1962, **9**, 297.
63. Cerrai, E., *Chromatographic Revs.*, 1964, **6**, 129.
64. Mould, D. L., Synge, R. L. M., *Biochem. J.*, 1954, **58**, 571.
65. Porath, J., Flodin, P., *Nature*, 1959, **183**, 1657.
66. Determann, H., *Angew. Chemie, Int. Edn.*, 1964, **3**, 608.
67. Moore, J. C., *J. Polymer Sci.*, 1964, **A2**, 835.
68. Porath, J., *Pure and Appl. Chem.*, 1963, **6**, 233.
69. Tiselius, A., Porath, J., Albertsson, P. A., *Science*, 1963, **141**, 13.
70. Gelotte, B., *New Biochemical Separations*, Editors James, A. T., Morris, L. J., van Nostrand, New York, 1964, p. 93.
71. Granath, K., *ibid.*, p. 110.
72. Laurent, T. C., Killander, J., *J. Chromatog.*, 1961, **5**, 103.
73. Lea, D. J., Schon, E. H., *Canad. J. Chem.*, 1962, **40**, 159.
74. Hjerten, S., *Archiv. Biochem. Biophys.*, Suppl. 1., 1962, p. 147.
75. Gelotte, B., Porath, J., *Chromatography*, Ed. Heftmann, E., Rheinhold, New York, 2nd Edition, 1967.

76. Joustra, M. K., *Protides Biol. Fluids*, 1967, **14**, 533.
77. Flodin, P., Killander, J., *Biochem. Biophys. Acta*, 1962, **63**, 403.
78. Gelotte, B., *Acta Chem. Scand.*, 1964, **18**, 1283.
79. Bengtsson, S., Philipson, L., *Biochem. Biophys. Acta*, 1964, **79**, 399.
80. Mulda, J. L., Buytenhuys, F. A., *J. Chromatog.*, 1970, **51**, 459.
81. Kalab, M., Martin, W. G., *ibid.*, 1968, **35**, 230.
82. Batlle, A. M. C., *ibid.*, 1967, **28**, 82.
83. Andrews, P., *Biochem. J.*, 1965, **96**, 595.
84. Bombaugh, K. J., Dark, W. A., King, R. N., *J. Polymer Sci.*, 1968, **C21**, 131.
85. Ross, J. H., Casto, M. E., *ibid.*, 1968, **C21**, 143.
86. Le Page, M., Beau, R., de Vries, A. J., *ibid.*, 1968, **C21**, 119.
87. Porath, J., Linden, E. B., *Nature*, 1961, **191**, 69.
88. Schmidt, M., *Biochem. Biophys. Acta.*, 1962, **63**, 346.
89. Kawade, Y., Okamoto, T., Yamamoto, Y., *Biochem. Biophys. Res. Commun.*, 1963, **10**, 200.
90. Gondko, R., Schmidt, M., Leyko, W., *Biochem. Biophys. Acta*, 1964, **86**, 190.
91. Björling, C. O., William-Johnson, B., *Acta. Chem. Scand.*, 1963, **17**, 2638.
92. Mathews, C. K., Brown, F., Cohen, S. S., *J. Biol. Chem.*, 1964, **9**, 2957.
93. Cuatrecasas, P., Anfinsen, C. B., *Ann. Rev. Biochem.*, 1971, **40**, 259.
94. Friedberg, F., *Chromat. Rev.*, 1971, **14**, 121.
95. Cuatrecasas, P., *J. Biol. Chem.*, 1970, **245**, 3059.
96. Axen, R., Porath, J., Emback, S., *Nature*, 1967, **214**, 1302.
97. van Deemter, J. J., Zuiderweg, F. J., Klinkenberg, A., *Chem. Eng. Sci.*, 1956, **5**, 271.
98. Huber, J. F. K. in *Comprehensive Analytical Chemistry Vol. IIB*, Eds. Wilson, C. L. and Wilson, D. W., Elsevier, 1968 pp. 1–53.
99. Perry, S. G., Amos, R., Brewer, P. I., *Practical Liquid Chromatography*, Plenum Press, New York, 1972.
100. Perry, S. G., *Chem. Brit.*, 1971, **7**, 366.
101. Snyder, L. R., *Gas Chromatography 1970*, Ed. Stock, R., Institute of Petroleum, London 1971, p. 81.
102. Kirkland, J. J. (Ed.), *Modern Practice of Liquid Chromatography*, Wiley-Interscience, New York, 1971.

103. *Gas and Liquid Chromatography Abstracts*, Ed. Knapman, C. E. H., 1970–72, Institute of Petroleum, London; 1973, Applied Science Publishers, London.
104. Conlon, R. D., *Analyt. Chem.*, 1969, **41**, 107A.
105. Zlatkis, A., (Ed), *Advances in Chromatography 1969*, Preston Technical Abstract Company, Evanston, Illinois, 1969.
106. Claesson, S., *Arkiv Kemi, Min. Geol.*, 1946, **23A**, 1.
107. Spackman, D. H., Stein, W. H., Moore, S., *Analyt. Chem.*, 1958, **30**, 1190.
108. Blaedel, W. J., Todd, J. W. *ibid.*, 1821.
109. Johansson, G., Karrmann, K. J., Norman, A., *ibid.* 1397.
110. Lovatt, S., Wright, W. C., *Chem. & Ind.*, 1961, 1433.
111. James, A. T., Ravenhill, J. R., Scott, R. P. W., *Gas Chromatography 1964*, Ed. Goldup, A. Institute of Petroleum, London, 1964, p. 197.
112. Horvath, C., Lipsky, S. R., *J. Chrom. Sci.*, 1969, **7**, 109.
113. Majors, R. E., *International Laboratory 1972*, p. 25.
114. Meiris, R. B., *Chem. & Ind.*, 1973, 642.
115. Loev, B., Goodman, M. M., *Chem. & Ind.*, 1967, 2026.
116. Bohen, J. M., Jouillé, M. M., Kaplan, F. A., Loev, B., *J. Chem. Ed.*, 1973, **50**, 367.
117. Brimley, R. C., Barrett, F. C., *Practical Chromatography*, Chapman and Hall, London, 1953, p. 94.

Chapter Three

Paper Chromatography;
Zone Electrophoresis

Paper Chromatography

Origin

Various types of simple separation on paper have been described as fore-runners of paper chromatography, among them a method of Rünge in 1850, and the process called 'capillary analysis' (Goppelsröder, 1909). Such methods were really more like adsorption chromatography, and paper chromatography as it is now understood was a development of the partition system introduced by Martin and Synge in 1941 (see Chapter 2). One of the solids which can be used to support the stationary phase is cellulose powder (p. 38). Consden, Gordon and Martin[1] in 1944 separated the amino-acids and peptides in wool protein hydrolysates by a method in which the powder column was replaced by a sheet or strip of paper, freely suspended in a vapour-tight vessel.

This is usually regarded as a typical partition system, where the stationary phase is water, held by adsorption on cellulose molecules, which in turn are kept in a fixed position by the fibrous structure of the paper. It is now realised, however, that adsorption of components of the mobile phase and of solutes, and ion-exchange effects, also play a part, and that the role of the paper is by no means merely that of an inert support.

The technique devised by Consden, Gordon and Martin has not undergone any fundamental changes, but there has been considerable improvement in detail, resulting partly from the wide variety of commercial apparatus which is available. Elaborate or expensive equipment is not essential, however, and very good results can be obtained with quite simple apparatus and materials.

Paper chromatography was originally used to separate mixtures of organic substances, but it was quickly adapted for the separation of inorganic ions, notably by Burstall[2], and by Pollard[3, 4] and their co-workers. Both anions and cations can be separated, and some simple quantitative methods have been developed.

One characteristic feature of the paper technique is that there is nothing corresponding to the column effluent of the usual gas or liquid systems. The separated substances are located and identified on the paper, and there is thus a reasonably permanent record of the separation. There is no collecting of a series of fractions, and sophisticated continuous monitoring devices are not needed. Quantitative determination of the separated substances may be carried out on the paper, but if it is desired to remove them all that is necessary is to cut out the part of the paper containing each substance and to elute each one separately.

General outline of the procedure

A drop of a solution containing the mixture to be separated is placed in a marked position on a sheet or strip of filter paper, where it spreads out to form a circular spot (Fig. 3.1 (*a*)). When the spot has dried the paper is put in a suitable closed apparatus with one end immersed in the solvent chosen as the mobile phase (but with the applied spot not covered, since the mixture would otherwise dissolve off the paper). The solvent percolates through the fibres of the paper by capillary action, and moves the components of the mixture to different extents in the direction of flow. It is

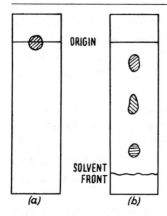

Fig. 3.1. Strip chromatogram

most important that the surface of the paper shall not become flooded with solvent, as there would either be no separation at all, or the zones would become very diffuse. When the solvent front has moved a suitable distance, or after a pre-determined time, the paper is removed from the apparatus, the position of the solvent front is marked, and the sheet is allowed to dry (Fig. 3.1 (*b*)). If the substances are coloured they are now visible as separate zones or spots; the object is so to choose the experimental conditions that the spots are as compact and as much separated as possible. If the substances are colourless they must be detected by physical or chemical means. The usual procedure is to apply a reagent or reagents, referred to as the *locating reagent*, which give a colour with some or all of the substances. Examination by ultra-violet light, or radio-chemical techniques (where applicable) can also be used.

When the positions of the separated zones have thus been detected, it is necessary to identify each individual substance. In the ideal case each one gives a characteristic colour with the locating reagent, as is often found with inorganic substances, but less commonly with organic. Treatment with a second reagent may cause characteristic changes in the colour of some of the spots, and disappearance of others (multiple dipping, p. 152).

The simplest method of identification is based on the R_f value, the ratio of the distance moved by the spot and that moved by the solvent front; it is discussed in detail on page 153. Direct comparison of R_f values is complicated by a number of factors, but a simple way of using the actual position of a spot to identify the substance is either to run several strips side by side in the same apparatus, or, better still, to use a sheet on which are placed spots of known reference substances, in addition to the mixture (Fig. 3.2 (*a*)). After development the spots appear as in Fig. 3.2 (*b*), from which it is clear that substances *A*, *B*, and *C*, but not *D*, are present in the mixture.

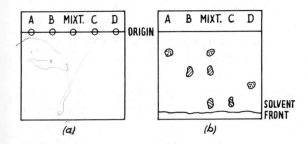

Fig. 3.2. One-way sheet chromatograms

Sometimes, particularly in the case of large groups of substances of similar chemical constitution, such as the amino-acids, the R_f values are too close together to give a good separation, whatever the solvent. In such cases the two-way technique devised by Martin is applied. A square sheet of paper is used, and the sample is placed at one corner (Fig. 3.3). Development with a solvent mixture 1, with the edge of the paper *AB* immersed, gives a partial separation similar to that obtained on a strip (Fig. 3.3 (*a*)). The sheet is removed and dried, and then put in a second apparatus with

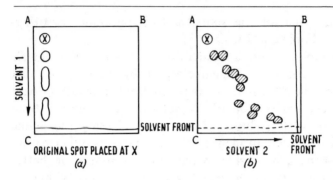

Fig. 3.3. Two-way chromatogram

the edge AC immersed in solvent 2. Development with this solvent, followed by drying and treatment with the locating reagent, gives a pattern of spots as in Fig. 3.3 (*b*). Apart from characteristic colour reactions, identification here is usually by comparison with a standard 'map' obtained by chromatography of known individual substances and mixtures.

The paper can be supported so that the solvent flows upward, downward, or horizontally. Results from the first and second methods are similar, but those from the third may be different. In the ascending method the paper strip is suspended vertically so that its lower end is just below the surface of the solvent, which ascends through the fibres of the paper by capillary action. There is therefore a theoretical limit to the height to which the solvent front can ascend. In practice this limit is not reached because too large a tank would be needed, and because the rate of flow becomes unacceptably slow towards the end of a long run. If the substances to be separated are of low or very close R_f values the length of run afforded by the ascending method in any apparatus of reasonable size may not be enough to give a good separation. This limitation is overcome by use of the descending method, in which the upper end of the paper strip is immersed in a suspended trough containing the solvent and the lower end hangs free (Fig. 3.4). The flow, although started by capillary action, is continued by gravity, and the distance moved by the solvent front is regulated only by the length of the paper and the height of the container. If the substances to be separated are of low R_f value they can be spread along the whole length of the strip by allowing the solvent to run off the end.

The horizontal method is fundamentally different. The sample spot is put at or near the centre of a horizontally supported piece of paper (usually a filter paper circle), and solvent is fed in at the centre. The flow is again by capillary action, but because the substances in the mixture spread out radially with the solvent, the separation has some distinctive features (p. 118).

When it is proposed to attempt a separation by paper chromatography the following points have to be considered:

(*a*) method (ascending, descending, or horizontal),
(*b*) type of paper,
(*c*) selection and preparation of the solvent (mobile phase),
(*d*) equilibrium in the selected apparatus,
(*e*) preparation of the sample,
(*f*) time of running (development), and
(*g*) method of location and identification.

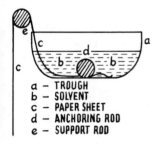

a — TROUGH
b — SOLVENT
c — PAPER SHEET
d — ANCHORING ROD
e — SUPPORT ROD

Fig. 3.4. Solvent trough for descending chromatography

Apart from the nature of the paper and of the solvent, the main factors which influence the separation are temperature, the size of the apparatus, the running time and the direction of solvent flow. The ultimate success of the experiment depends on the efficacy of the detection procedure. A detailed investigation of the effect of all these factors on the separation of amino-acids was carried out by Hanes and his co-workers[5, 6, 7], and also by Knight[47].

Apparatus and techniques
Descending method

The original apparatus of Martin was simple, and, as is often the case with new techniques, was constructed to some extent by adapting materials intended for other purposes. The basic requirement is a vessel or 'tank' with a well-fitting lid to prevent the escape of solvent vapours. Near the top of the tank are supports for a trough in which the solvent is placed. The end of the paper in the trough is held down by a glass rod or strip, which suffices to hold the paper in position even when it is saturated with solvent. To ensure that the solvent flows regularly and evenly without flooding, the paper passes over a glass rod supported parallel to the edge of the trough (Fig. 3.4). For the first few centimetres flow is by capillary action, gravitational flow only beginning when the front has passed over the glass rod.

Tanks used for paper sheets are usually rectangular but cylindrical tanks are convenient for paper strips. The main difference in design of the various equipments is in the method of supporting the trough. In some proprietary apparatus polythene or stainless steel brackets, bolted through the sides, are used; in others a polythene or stainless steel frame stands on the bottom of the tank; the lower part holds a dish in which solvent can be put to saturate the atmosphere, and the top holds the solvent trough; in a third method the whole trough and glass rod system is fixed to the lid, and is lifted out with it as a single unit.

Ascending method

The same kind of tank is used for ascending as for descending chromatography, but the solvent is simply put in the bottom, and the paper is suspended above it. One method is shown in Fig. 3.5 (*a*), where a sheet is hung on a glass rod, which may be held in fixed brackets, or may be jammed in by means of pieces of rubber or polythene tubing on its ends. The paper is sewn to form a loop over the rod, or is clipped on. The simplest method requires no paper support at all. The sheet is formed into a cylinder, with the samples on a circumferential line near the bottom. The meeting edges are clipped or sewn together, and the cylinder stands with its lower edge in the solvent (Fig. 3.5 (*b*)). An elegant method of this type, due to Hunt, North, and Wells[8], is illustrated in Fig. 3.6. The paper is in the form of a rectangular sheet (CRL/1), with a series of parallel slits cut in it. This is a small-scale method, the sheets being only.21.0 x 11.0 cm, and the slits allow an adequate number of spots to be put on without sideways diffusion into each other. The effect is thus of a set of

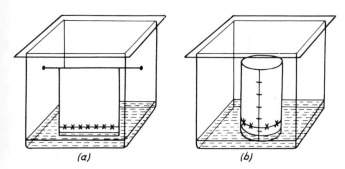

(a)   *(b)*

Fig. 3.5. Arrangements for ascending chromatography with paper sheets

separate strip chromatograms. These papers are known as 'Chinese lanterns', from the shape they assume when the solvent run is complete. A suitable tank is a one dm³ tall form beaker, without pouring spout, covered with a clock-glass (Fig. 3.6 (*b*)).

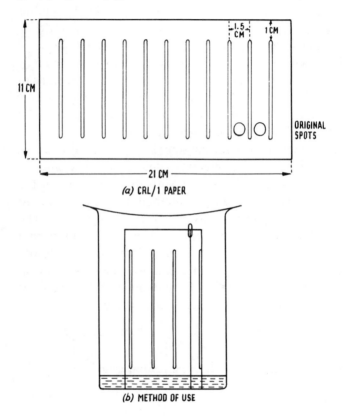

(a) CRL/1 PAPER

(b) METHOD OF USE

Fig. 3.6. Small slotted sheets for ascending chromatography

The mechanical weakness of the wet paper limits the size of sheet which can be used in this way. Omission of the slits allows a bigger sheet to be used, but requires larger spacing of the samples. If the sample is put at one corner the method can be used for two-way separations, the cylinder being reformed along an axis at right angles to the first for the run in the second solvent.

Fig. 3.7 shows various ways in which an ordinary gas jar can be used for making strip chromatograms. In the method of Fig. 3.7 (*a*), a cork with a glass hook passing through it is used as the closure; Fig. 3.7 (*b*) shows a system with a larger glass jar, where the cork is placed in a small hole at the centre of a flat glass disc; in a gas jar 10 cm in diameter this method allows a triple hook for three strips to be used (Fig. 3.7 (*c*)); Fig. 3.7 (*d*) illustrates an arrangement of glass plates[9], which are held together with rubber bands. In all the methods of using gas jars the paper can be put in and allowed to hang above the solvent to attain equilibrium with the atmosphere in the jar. The end of the strip is then lowered into the solvent by pushing the glass hook through the cork, or by pushing the strip itself through the slit (Fig. 3.7 (*d*)). In the last type of apparatus the atmosphere can be

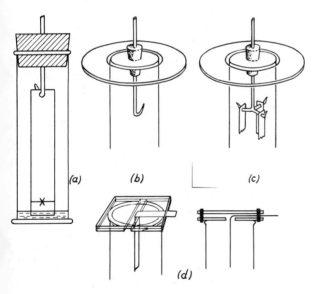

Fig. 3.7. Arrangement of gas jars for ascending chromatography

allowed to equilibrate, and the strip can then be inserted through the slit without disturbance.

If many separations have to be carried out it is usual to use sheets, with several samples on each, rather than a number of strips. To increase the number of sheets which can be put in a descending apparatus, arrangements such as those shown in Fig. 3.8 can be used. Type (*a*) is quite satisfactory, although there is some danger of the two middle sheets touching when wet with solvent, particularly when they are being removed from the tank. Not more than two troughs should be put in a tank of about thirty cm in width. Type (*b*) is not as

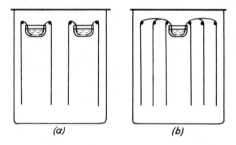

Fig. 3.8. *Arrangement of several sheets in one tank for descending chromatography*

satisfactory; inserting and removing the papers is difficult, solvent flow to the outer sheets is less regular, and the full area of the outer sheets is not used. The simplest way of arranging several ascending sheets in one tank is illustrated in Fig. 3.9.

If two wet sheets touch each other, or if solvent is splashed on to a sheet in removing it from the tank after the run, the sheets concerned must be regarded as failures. To reduce such difficulties, and to give greater ease of handling, a frame was designed by Dent[10] (developed by Smith[11], and others) to hold a number of square sheets. The frame consists of two rigid plates joined by four rods passing

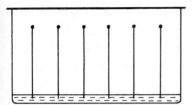

Fig. 3.9. Arrangement of several sheets in one tank for ascending chromatography

through holes at the corners of the paper sheets (Fig. 3.10). The papers are separated by collars, and when the end nuts are screwed up the sheets are held fairly tautly. Special square sheets of paper with corner holes can be obtained for these frames (25 cm square being the common size), but in a later design larger polythene spacers are used, which simply grip the corners of the sheets. Frames (with the appropriate tanks) to take from five up to fifteen sheets are made.

Samples can be applied when the sheets are assembled in the frame. For two-way separations the arrangement is ideal, as, once loaded in the frame, the papers need not be handled

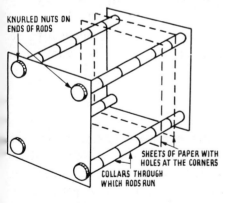

KNURLED NUTS ON
ENDS OF RODS

SHEETS OF PAPER WITH
HOLES AT THE CORNERS

COLLARS THROUGH
WHICH RODS RUN

Fig. 3.10. Frame for holding square sheets of paper

until the two runs are finished. After the first run the frame is simply lifted out and left for the solvent to dry off completely. It is then turned through 90° and placed in the second tank. Some workers have experienced trouble through sagging of the wet sheets causing them to touch, and a tensioning system to overcome this has been designed[12].

Originally, particularly for two-way separations, large sheets up to 45 x 55 cm were used. Although the tendency now is towards the use of smaller sheets, tanks to take the large sizes are still made; they are all-glass tanks of the 'fish-tank' type, wood and glass cabinets, or stoneware vessels. The advantage of the last seems to be better thermal insulation, giving less temperature variation inside. The smaller tanks of cubic, rectangular and cylindrical shapes are customarily all-glass. Solvent troughs for descending chromatography are in most cases of glass, although plastics can be used if they are not attacked by the solvent. Stainless steel is also used, but it is not suitable for solvents containing mineral acids, particularly hydrochloric. Polythene frames, troughs, and fittings seem to be least affected by the usual chromatographic solvents, and are preferable to much of the rather delicate glass apparatus. Some manufacturers produce 'universal outfits', in which one tank is supplied with a range of internal fittings and ancillary apparatus to allow it to be used for ascending or descending one- or two-way chromatography at will.

Horizontal chromatography

The technique of horizontal chromatography, sometimes known as the Rutter method[13], involves a slightly different principle from those so far described. The sample spot is placed at or near the centre of a piece of filter paper (usually circular), which is held horizontally; solvent is applied at the centre, from which it spreads out radially. The substances are thus spread into a series of concentric bands; whereas in the upward and downward methods the shape of the original spot is roughly maintained during migration, in this case the shape of the spot is lost (Fig. 3.11). There is an advantage here, since there is a concentration effect as the annular zones are formed, and a better separation may result than in the other methods.

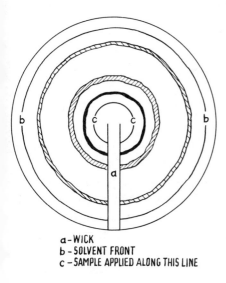

a – WICK
b – SOLVENT FRONT
c – SAMPLE APPLIED ALONG THIS LINE

Fig. 3.11. Horizontal radial chromatogram

A number of arrangements have been described for this technique; most of them require only very simple apparatus, the main differences being in the way of applying the solvent. In the simplest, and in many ways the most satisfactory, method, a radial strip about 4 mm wide is cut to form a wick (Fig. 3.11). The paper is sandwiched between the edges of two Petri dishes, so that the wick dips into solvent contained to a depth of about 5 mm in the lower one (Fig. 3.12 (*a*)). The sample is applied at the centre. The run is continued until the solvent front reaches the edge of the dishes. In a

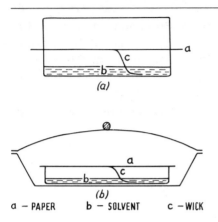

(a)

(b)

a — PAPER   b — SOLVENT   c — WICK

Fig. 3.12. Arrangements for horizontal chromatography

similar arrangement (Proom and Woiwod[14]) the Petri dish is put in a larger vapour-tight container (Fig. 3.12 (*b*)). Another method of arranging the solvent flow is to make, in the centre of the paper, a small hole through which passes a wick made from a small piece of paper, or a cotton thread (Fig. 3.13 (*a*)). The sample spots are applied at intervals on a line round the centre (Fig. 3.13 (*b*)).

In both methods the effectiveness of the separation, depending as it does on avoiding flooding of the paper, is controlled by the size of the wick, which in turn will depend on the type of paper and the solvent. Some experiment is needed in each case to find the best conditions, and even then great care is needed in forming the wick if reproducible results are to be achieved.

An attempt to remove this difficulty, and to give a constant solvent flow, is found in an apparatus designed by Kawerau[15]. The wick is replaced by a capillary tube with a ground-glass top (adjustable for height) on which the centre of the paper rests. The tube is supported in a 'beehive shelf' standing in an annular groove in the lower part of a glass container shaped like a desiccator (Fig. 3.14 (*a*)). Special

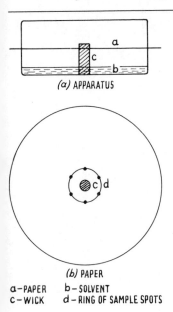

(a) APPARATUS

(b) PAPER

a – PAPER   b – SOLVENT
c – WICK   d – RING OF SAMPLE SPOTS

Fig. 3.13. Arrangements for horizontal chromatography

papers with five radial slits are used (KCT 14, Fig. 3.14 (*b*)), and they are kept in place by the lid. The presence of the slits helps to homogenise the atmosphere in the container, and also keeps the migrating zones separate from each other. A further advantage of the Kawerau apparatus is that it provides support for the centre of the paper.

The main advantages of horizontal methods lie in the simplicity of the apparatus, and in economy in paper and solvents. Except in the Kawerau apparatus, ordinary filter paper discs are used, with only a few cubic centimetres of solvent. There is also a great saving in time. With a fast paper and a free-flowing solvent, on an 11 cm disc, a run may be completed in thirty to sixty minutes. Differential treatment of the developed zones is easy, and the concentration effect is often useful.

Some workers have considered that this method is not

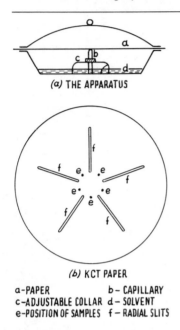

(a) THE APPARATUS

(b) KCT PAPER

a–PAPER   b– CAPILLARY
c–ADJUSTABLE COLLAR d– SOLVENT
e–POSITION OF SAMPLES f – RADIAL SLITS

Fig. 3.14. Kawerau apparatus

quick enough, and apparatus has been designed to give a centrifugal acceleration to the solvent flow by rotating the paper[16, 17]. The centrifugal method has been found useful in the separation of compounds labelled with radioactive isotopes with short half-lives[18].

Choice of method

In deciding which method (ascending, descending, or horizontal) is to be used for a particular experiment, the personal preference of the worker and the equipment available are probably the primary considerations. There is little difference in the results of upward and downward chromatography, but the time taken varies. For research problems this may not be an important consideration, but for routine analyses it is probably the deciding factor. If a run cannot be conveniently completed in one day of, say, eight hours, it is better to use overnight running, with a run of fifteen hours or more. The time taken depends mainly on the solvent and the paper, but for a particular solvent/paper combination downward running is quicker than upward. The time of flow of a downward chromatogram can be extended by trimming the strip to give a constriction between the reservoir and the position of the sample spot (Fig. 3.15).

For a preliminary investigation, where a number of different solvent mixtures and, perhaps, different papers, are to be tested for a particular separation, the gas jar (ascending), the slotted paper sheets or the horizontal methods are

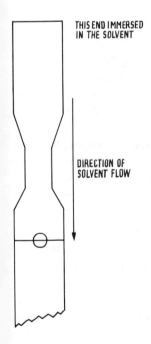

THIS END IMMERSED
IN THE SOLVENT

DIRECTION OF
SOLVENT FLOW

Fig. 3.15. Constriction of the paper strip to reduce the rate of solvent flow

most convenient, and since they need the simplest apparatus it is not difficult to do many runs simultaneously.

For two-way separations, or for simultaneous one-way (ascending) separation of a large number of different samples with the same solvent, the square frame type of apparatus is most convenient. Only if the short length of run (25 cm) is regarded as insufficient is it necessary to use larger sheets and tanks; if those are used, the descending method will be found to be the better.

For quantitative separations, or those where a long equilibration is to be allowed, the descending method seems to be favoured; it is also best if a very long run is required, as the solvent can run off the paper.

Paper

The original work in paper chromatography was done on Whatman No. 1 filter paper, and this, in its modern form, is probably still the most widely used; it would usually be the first choice if a new separation were to be investigated although some workers have preferred No. 3[5]. At first it was necessary to use the ordinary grades of paper, but the makers now supply products specially made for chromatography, in which the chemical and physical properties of the material are closely controlled in manufacture to give papers of high purity and uniformity.

Paper for chromatography is made from cotton cellulose, and since this is a natural product there will always tend to be traces of organic and inorganic impurities in it. The residual levels of trace metals and 'ash' ($5-10 \mu g \, cm^{-2}$, depending on thickness) do not interfere with most separations. Cotton waxes and similar substances are largely removed during manufacture, and attempts to remove the last traces may degrade the cellulose. Nevertheless, from time to time various workers have recommended that papers should be washed with solvents before use, and some have even designed special apparatus for the purpose (although any descending-type apparatus will serve). Pre-treatment to remove impurities may, however, destroy some of the desirable properties of the paper, and it is recommended that it should be avoided unless the circumstances are exceptional.

Each manufacturer supplies chromatography paper in several different grades, the differences being principally in the density and thickness. The grade chosen for a particular separation will depend on the loading and rate of flow required. The primary role of the paper is to act as a support for the stationary phase. In this role the chemical nature of the paper is probably of first importance, and there is little, if any, chemical difference between the various grades of pure cellulose paper. The many free hydroxyl groups in the cellulose molecule render the surface hydrophilic, and thus give it an affinity for the more polar solvent molecules. There must also be other adsorption effects, since the cellulose has a similar affinity for the more polar solutes as well, and there will usually be ion-exchange effects due to carboxyl groups, of which a few are always present in cellulose molecules.

The rate of flow of the mobile phase depends on the viscosity of that phase, and for a given solvent mixture the rate depends on the physical nature of the paper. Paper consists of a mass of small cellulose fibres randomly matted together to form a three-dimensional network with relatively large open spaces. The stationary phase is adsorbed on the surface of the fibres; the mobile phase flows over the surface and the adsorbed layer and fills the spaces. If the fibres are more closely packed, to give an increased density, the area of free surface and the size of the spaces are reduced, and the flowrate decreases. Conversely, making the paper thicker without changing the density tends to increase the flowrate. (Flowrate in this context is the rate of movement of the solvent front, and not the volume flowing in unit time). There is, however, no simple relationship between density, thickness, and flowrate, because the various grades of paper have slightly different fibre characteristics 'built-in' during manufacture.

In choosing a paper one must make a compromise between maximum efficiency and available time. Resolution (the extent to which individual substances are separated from each other) is greatest with a low flowrate; that is, on a slow paper. If optical or radio-active scanning is to be used for location (p. 164) a smooth slow paper is recommended. For separation of larger amounts of material, or for preparative separations, a thicker paper is required, so that the higher loading can be achieved without increasing the area of the original spot. Efficiency of separation is very markedly reduced by excessive loading of thin papers, but the thicker papers accommodate the higher loading without trouble.

The characteristics of Whatman chromatography papers are shown in Table 3.1. Of the papers listed, Nos. 1 and 4 are the most useful for general chromatography, No. 4 in particular when higher speed is required. Nos. 3MM, 17, and 31ET are more often used for paper electrophoresis.

Table 3.1   *Characteristics of Whatman chromatography papers*

Grade	Flowrate water mm/30 min	Density g cm^{-3}	Thickness mm
20	85	0.58	0.16
2	115	0.54	0.18
1	130	0.54	0.16
3MM	130	0.56	0.33
3	130	0.49	0.38
4	180	0.46	0.20
17	190	0.50	0.88
31ET	225	0.36	0.53

Paper is supplied in various standard sheets, circles, and reels, and in special shapes. It should be stored away from any source of fumes (especially ammonia, which has a high affinity for cellulose), and should not be subjected to large changes in humidity. Reels of smaller width are conveniently kept in some form of container or dispenser (Fig. 3.16) which allows the required length to be pulled out and cut off. These reels are wound on a cardboard core. It is best to mark the type number on the core as soon as the reel is unwrapped, otherwise it may not be easy to identify the paper later on.

Because of the method of manufacture, paper has a slight 'grain', known as the machine direction, and separations may differ according to whether they are made parallel to or perpendicular to this 'grain'. Sheets to be compared with each other should as far as possible be run with the same orientation, but the effect is not usually important. In rectangular sheets the machine direction is parallel to the longer side, but with sheets supplied as squares it is not easy to tell which direction is which. The machine direction is usually marked on the packet in which the paper is supplied.

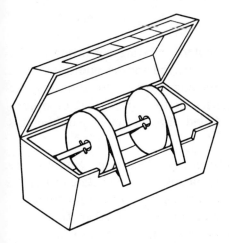

Fig. 3.16. Container/dispenser for reels of paper

There are several ways in which paper can be modified for chromatography. It can be treated with a hydrophobic substance so that it will support a non-aqueous stationary phase ('reversed phase'); it can be impregnated with a substance which will complex with one or more of the substances being separated, it can be given ion-exchange properties by combination with a resin or by chemical modification (pp. 157 and 158), or it can be impregnated with an adsorbent such as silica gel so that it can be used for adsorption chromatography (p. 311).

It is in principle an attractive idea that a single standard paper should be developed, with properties chosen to give the best resolution, control of flowrate being by means of wicks of other types of paper inserted between the start-line and the solvent reservoir. Knight[47] published some preliminary work on the use of such a system, but this was not followed up, probably because of difficulty in choosing the right compromise properties for the standard paper. It is therefore still necessary to make a choice of paper before starting a separation.

Solvents

The mobile solvent is normally a mixture consisting of one main organic component, water, and various additions, such as acids, bases or complexing agents, to improve the solubility of some substances or to depress that of others. Anti-oxidants may be included. The principal requirements for a solvent are:

It should be reasonably cheap, since large amounts are often used. It must be obtainable in a good state of purity. It is now possible to buy pure solvents for chromatography which for most purposes need no further treatment. Mixtures of isomers, such as xylene, pyridine homologues, or petrol, tend to be of variable composition, and are best avoided. Traces of metals are undesirable, even for organic separations. The solvent should not be too volatile, because of the necessity for more meticulous equilibration; on the other hand, high volatility makes for easy removal of the solvent from the sheet after the run. Its rate of flow should not be greatly affected by changes in temperature.

The reasoning which can be used in solvent selection may be illustrated by the following examples. For polar organic substances more soluble in water than in organic liquids, there will be little movement if an anhydrous mobile phase is used; adding water to the solvent will cause those substances to migrate. Thus butan-1-ol is not a solvent for amino-acids unless it is saturated with water (a property used in the old method of separating amino-acids into various groups); addition of acetic acid allows more water to be incorporated, and hence increases the solubility of amino-acids, particularly basic ones; this three-component mixture is a very good one for separating amino-acids. Many other polar substances with solubility characteristics similar to those of amino-acids, such as indoles, guanidines and phenols, can be separated with this mixture. Inorganic ions are usually separated as complex ions or chelates with some solubility in organic solvents; for instance, iron forms a complex chloride ion which is very soluble in aqueous acetone, whereas nickel does not so readily form such an ion; iron and nickel can therefore be separated with this solvent, so long as hydrochloric acid is present to stabilise the complex ion.

Similar arguments can be applied to most separations, but the choice of solvent for a particular purpose is still to a large extent empirical. So many separations have been reported, however, that it is not usually difficult to find a suitable one to use as a starting point for new work, but a knowledge of the chemical properties of the substances to be separated is clearly desirable.

Some examples of the types of solvent mixture which have been used in various representative separations are shown in Table 3.2. They are given simply as examples, and are not necessarily the best for the separations quoted, although they are all perfectly satisfactory; it is not really possible to recommend a 'best' solvent for any particular purpose, as the views and requirements of authors vary considerably, even when they are working in closely related fields.

Some of the solvents, when mixed in the proportions shown, give a two-phase system. It is the custom in such cases, when settlement into two layers is complete, to use the aqueous layer to saturate the atmosphere by putting it in a suitable vessel in the bottom of the tank; this is simply an economy measure, since both layers will be in equilibrium with vapour of the same composition. The organic layer, which must be absolutely homogeneous, is put in the solvent trough. I. Smith[11] examined the organic layer in several instances, and showed that good results can be obtained by making up mixtures having nearly the composition of these layers. The advantages of this are: preparation of the solvent can be done by mixing in a measuring cylinder — there is no waiting for separation into layers (a lengthy process with solvents such as phenol/water); the exact composition of the solvent is known; and as the mixture is not saturated at room temperature it will not separate into two phases if the temperature varies within reasonable limits (say ±5°C) during the run.

For reproducible results the solvent mixture must be made up with care, although only to the accuracy given by a measuring cylinder. Solvents should not be over-worked. For overnight runs the solvent should be used once only. Solvents used in small vessels for short runs may be used several times for successive runs, but they should not be kept for more than one day; very volatile solvents are best mixed afresh each time they are required.

Table 3.2 *Solvents for paper chromatography*

Separation	Solvent	Proportions	Reference
Amino-acids	phenol/water	Sat. soln.	
	phenol/water	500 g/125 cm^3 *	11 (p. 84)
	phenol/water/ ammonia	Sat. soln. 200 : 1	
	butan-1-ol/acetic acid/water	4 : 1 : 5	26
	butan-1-ol/acetic acid/water	12 : 3 : 5*	11 (p. 84)
	butan-1-ol/pyridine/ water	1 : 1 : 1	27
Sugars	ethyl acetate/ pyridine/water	2 : 1 : 2	28, 29
	ethyl acetate/ pyridine/water	12 : 5 : 4*	11 (p. 248)
	ethyl acetate/ propan-1-ol/water	6 : 1 : 3	28, 29
	ethyl acetate/ propan-1-ol/water	14 : 2 : 4*	11 (p. 248)
	ethyl acetate/acetic acid/water	3 : 1 : 3	28, 29
	ethyl acetate/acetic acid/water	14 : 3 : 3	11
Fatty acids	butan-1-ol/1.5 mol dm^{-3} NH_3	Sat. soln.	30
Co, Mn, Ni, Cu, Fe (chlorides)	acetone/conc. HCl/ water	87 : 8 : 5	2
F, Cl. Br, I (Na salts)	pyridine/water	90 : 10	3
Hg, Pb, Cd, Cu, Bi (chlorides)	butan-1-ol/3 mol dm^{-3} HCl	Sat. soln.	2
As, Sb, Sn (chlorides)	pentan-2,4-dione (sat. soln. in water)/ acetone/conc. HCl	149 : 1 : 50	2

* Monophasic version of the preceding mixture.

Equilibrium

Study of the literature of paper chromatography reveals many views on the establishment of 'equilibrium' in the tank, and some authors have given it special attention[5, 19, 20, 50]. It is not possible to generalise because the degree of equilibration needed depends on the size of the apparatus, the solvent system and the nature and purpose of the separation. The object is to prevent evaporation of the solvent from the paper.

Since the basis of the paper chromatography process is distribution between cellulose-bound water and a moving organic solvent, it has been usual to choose as the mobile phase an organic liquid which is only partly miscible with water, and to saturate it with water before use. The simple picture of a water-solvent boundary as is seen in solvent extractions cannot exactly represent the situation on paper, because it has been found that solvents completely miscible with water can be used, and that partly miscible ones need not be saturated. It is probably best to consider that there is a layer of water molecules held on the cellulose by hydrogen bonding, and that this (the stationary phase) absorbs the solute molecules from the moving solution. There is thus no definite solvent-solvent interface, but there is a gradual change in composition from polar solvent (water) to the less polar organic solvent in the direction away from the surface of the cellulose molecules (Fig. 3.17). This is only possible if the layers concerned are fairly thin, and explains why an excess of either phase ('flooding') prevents separation by causing extensive diffusion.

As the mobile phase flows through the fibres of the paper

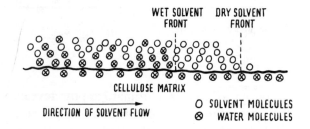

WET SOLVENT   DRY SOLVENT
FRONT   FRONT

CELLULOSE MATRIX

DIRECTION OF SOLVENT FLOW

O SOLVENT MOLECULES
⊗ WATER MOLECULES

Fig. 3.17. Schematic diagram of the solvent layers in a paper chromatogram

the cellulose will tend to adsorb more water from it. The result is that the moving solvent tends to become denuded of water as it advances, and its composition is not constant along the sheet[21] (Fig. 3.17). Sometimes there is a definite boundary where the solvent composition changes, for example, in some inorganic separations where acetone/water/hydrochloric acid mixtures are used as the solvent there is a 'dry' solvent front, and, some distance behind it, a 'wet' solvent front. The forward area consists of acetone from which the water has been removed, and the area behind the wet solvent front consists of aqueous acetone, and therefore contains all the acid. This effect is well shown in the experiment described on p. 325 (the existence of two fronts is not normally revealed until the locating reagent has been applied).

This purely descriptive analysis reveals a situation of great complexity. If the composition of the solvent varies continuously along the paper, the composition of the vapour with which it is in equilibrium also varies. As it is not possible to arrange a corresponding concentration gradient in the atmosphere in the tank, there may tend to be further changes in solvent composition due to evaporation of the more volatile constituents. A further complicating factor is the amount of water held by the paper initially. It is thus necessary to consider to what extent the original solvent mixture and the paper must be equilibrated with the atmosphere before the run starts. For most applications it seems that it is not necessary to go to extreme lengths, and some of the meticulous and elaborate equilibration carried out by earlier workers was probably not really necessary. Some authors recommend that for a particular separation (such as DNP-amino-acids[11]) the paper should be equilibrated for a long time with the solvent atmosphere. Others recommend, for example, in some inorganic separations (Pollard[9]), putting the paper into an already equilibrated tank and allowing the run to begin immediately.

The time taken for the atmosphere to reach equilibrium with the solvent depends on the size of the tank and the volatility of the solvent. If very volatile mixtures, such as those containing lower alcohols, ethers or ketones, are used, evaporation from the paper will be more rapid, and equilibration is important. Reducing the volume to be saturated with vapour will clearly make equilibration easier and more efficient, and it is a general rule that the tank should be of the smallest possible size to hold paper of the desired dimensions. This is partly responsible for the general reduction in size of apparatus to which reference has already been made. It is also the reason why the gas jar and horizontal methods, with their very small surrounding atmosphere, are so effective.

When chromatography is being carried out to determine physical values (such as when R_f is to be correlated with partition coefficients) or for quantitative determination of inorganic substances, or for some organic separations of particular delicacy (steroids and barbiturates, for example[11]) great care must be taken over equilibration. But for repetitive purposes in normal qualitative and semi-quantitative separations it is more important to standardise the experimental conditions (including the time of running) and the manipulative procedure.

A number of techniques can be adopted to ensure that the tank is filled with solvent vapour. The first essential is a well-fitting lid; if there is any doubt about its vapour-tightness, the edges of the tank should be smeared with vaseline or silicone grease, care being taken that none can get on to the paper or into the solvent. For downward separations, vessels containing the solvent mixture are put at the bottom of the tank, and the paper is inserted with its support and retaining rods, the solvent trough being empty. The dry paper is left in the tank while the atmosphere is equilibrated; one way of accelerating this process is to line the walls of the tank with paper soaked in the solvent. After the desired period, solvent is introduced into the trough with a pipette through a small hole in the lid, which is then closed with a bung or glass disc.

For upward separations it is not so easy to introduce the solvent after the paper. In gas-jar methods the paper suspension is arranged so that when the strip is first inserted it hangs above the solvent layer, and after the desired interval it can be lowered so that its end dips into the solvent (Fig. 3.7). If the solvent is swirled round the walls of the gas jar before the paper is put in, only a few minutes settling time need be allowed between insertion of the strip and the start of the run. If the arrangement of Fig. 3.7 (d) is used, the atmosphere can be equilibrated before the paper is inserted. In larger tanks fitted for the ascending method, it is simpler to omit any time for equilibration after the paper is put in. In the multi-sheet frame type of apparatus solvent can be put round the tray in which the frame stands, and the running solvent can be introduced subsequently with a long pipette through a hole in the lid. Alternatively the frame can be suspended above the solvent in the tray by means of a cord passing through the hole in the lid; it can then be lowered when required. For horizontal separations the very small volume of the tank makes equilibration unnecessary.

It will be observed that in most cases the paper is allowed to be in contact with the solvent vapours for a time before the run begins. During this period the atmosphere will approach saturation with the vapour, and it may be that after a certain time equilibrium in this respect is attained. It is, however, arguable whether the paper attains equilibrium with the atmosphere. The cellulose will tend to adsorb water from, or give it up to, the atmosphere, and it may also adsorb organic solvent constituents. In extreme cases the paper may adsorb so much that no clear solvent boundary can be seen during the chromatographic run. This leads to excessive diffusion of the solutes and inefficient separation. Probably all that is needed is a short period (say, up to one hour) in which the disturbance caused by the insertion of the paper can settle down, and in many cases even this short wait is superfluous. One should establish whether careful equilibration makes any important difference if all the other factors are standardised, and if the separation is reproducible and satisfactory without it, equilibration can well be omitted.

It might at first sight seem attractive to eliminate the influence of the atmosphere in the tank by sandwiching the paper between glass plates. This, however, usually causes uneven solvent flow because of irregular contact between paper and glass, and it is not often used, except in some separations of volatile substances; for example, the separation of peroxides (Cartlidge and Tipper[22]).

Associated with the problem of equilibrium is that of temperature control. It is not necessary to use any elaborate methods, but all chromatography should be carried out in a room which is not subject to major temperature changes, and where the tanks are protected from draughts and direct sunlight. If accurate control of temperature is wanted, one of the very small types of apparatus can be used in an oven or incubator. Special apparatus has been designed for this purpose, such as the *Chromatocoil* (Schwarz[23, 24]).

Changes in temperature can set up convection currents in the tank, giving local inhomogeneity in the atmosphere and causing changes in the relative rates of evaporation of solvent components from the paper; the partition coefficients of the various solutes will alter with temperature; and the viscosity of the solvent, and hence the rate of flow, may be markedly affected by temperature changes. The combined effect of these factors may make the separation virtually useless. If the solvent is a mixture saturated with water a reduction in temperature may cause it to separate into two phases; this will spoil the chromatogram, but can be avoided by using non-saturated solvent mixtures (p. 130). It must be emphasised that these difficulties apply to temperature changes during the run. It has been proposed that chromatography at controlled elevated temperatures (up to about 60°C) might be used to give quicker separations (but see p. 154).

If the solvent mixture contains substances which can react together it is sometimes suggested that the reaction should be allowed to go to completion before the mixture is used. For example, in a number of separations, notably for amino-acids, a mixture of butan-1-ol, acetic acid, and water is used. Some authors[25] suggest that the mixture should be left until it contains the equilibrium concentration of butyl acetate, but the reaction takes a matter of days at room temperature, and although the results do depend on the amount of ester present, a prefectly adequate separation is obtained if the mixture is used as soon as it has been made up.

Preparation of specimens

The mixture to be separated is applied to the paper as a solution. The nature of the solvent is immaterial, as long as it will evaporate completely without leaving a residue, and without attacking the paper. Solid samples, such as soils, or biological cell or tissue material, are macerated with the solvent, or submitted to some standard extraction procedure (such as Soxhlet). Many important samples, such as urine or other biological fluids, are already in an aqueous medium. In other cases water is used as the solvent when the substances are soluble in it.

The extraction procedure inevitably extracts more than just the substances to be tested for. Chromatography itself will remove many interfering substances, though not in all circumstances. Often, therefore, a preliminary separation is required. Since this depends both on the nature of the specimen and on the substances being looked for, it is not possible to generalise about methods. The most useful are electrolytic processes, ion-exchange, and solvent extraction.

It was noticed at an early stage that high concentrations of inorganic ions caused poor separation of amino-acids, and of some other types of substance. Aqueous biological extracts, urine, neutralised protein hydrolysates, and other solutions which may have to be examined for amino-acids and sugars will always contain appreciable amounts of inorganic material. Removal of these is called 'desalting'; it is important, and should always be carried out if it can be done without affecting the organic compounds. The earliest method was an electrolytic one, using a mercury cathode to remove metal ions. Two later processes are based on the use of ion-exchange materials; one is a column technique with the type of resin described in Chapter 2, and the other is electro-dialysis with an ion-exchange membrane. Solvent extraction is sometimes preferred for desalting non-ionic substances, in cases where undesirable losses or changes occur when the other methods are used. The choice of method depends on the apparatus available and on the electro-chemical nature of the substances to be separated. The electrolytic method seems to be the most generally useful and convenient.

Electrolytic desalting

The principle of the electrolytic method is illustrated in Fig. 3.18. The sample is in a cell bounded on one side by a flowing stream of mercury, which forms the cathode, and on the other by a dialysing membrane separating it from a stream of aqueous acid, in which the anode is placed. When the current is flowing, the cations are discharged at the mercury surface. The resulting amalgam flows through a water cell, in which it is decomposed, the mercury being recycled. The anions from the sample pass through the membrane into the acid stream, which is allowed to flow to

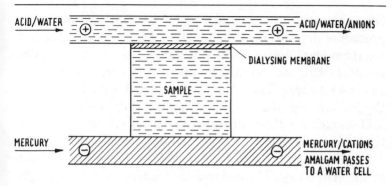

ACID/WATER → ⊕ ⊕ ACID/WATER/ANIONS →

DIALYSING MEMBRANE

SAMPLE

MERCURY → ⊖ ⊖ MERCURY/CATIONS
AMALGAM PASSES
TO A WATER CELL

Fig. 3.18. Principle of electrolytic desalting

waste. Solutions containing much ammonium ion cannot be desalted in this way, because the amalgam is not decomposed by water; it sticks to the glass and impedes the mercury circulation. A typical desalting experiment would need about 2 cm³ of solution, and an applied voltage of 200 V d.c. for about fifteen minutes. The current would be about 0.8 A initially, falling at the end quite rapidly to about 0.1 A.

Desalting must be carried out with care. Very small amounts of metal ions are not detrimental, and attempts to remove the last traces may result in decomposition of some amino-acids, and in the complete removal of others; for example, arginine may be converted into ornithine. Other chemical changes are due to the reducing atmosphere in the cell; thus α-oxo-acids may be reduced to α-hydroxy-acids; diiodotyrosine may undergo hydrogenolysis to tyrosine. Proteins and some amino-acids may be precipitated in the cell; anions of some strong organic acids and some small neutral molecules, such as monosaccharides, may be lost by diffusion through the membrane.

Ion-exchange desalting

The type of column which can be used is shown in Fig. 2.7. A column 10 cm long and 1.0–1.5 cm in diameter is adequate, since the volumes to be desalted are small. The procedure depends on whether the substances being desalted are themselves ionic or not. For instance, amino-acids can be desalted with Zeo-Karb 225 or an equivalent strong acid cation-exchanger. The column is prepared with the resin in the H-form, and the solution of amino-acids (which must be acid enough for them all to be present as cations) is put on; the amino-acids and the other cations are adsorbed on the resin, displacing H^+, and the anions flow through the column in the acid eluate. The amino-acids (except the most strongly basic) are eluted with aqueous ammonia, other cations remaining on the column.

During the desalting of sugars, or other neutral molecules, the cation-exchange resin removes the cations, but the initial eluate contains the neutral molecules and the anions, and is strongly acid. The anions can be removed by subsequently passing the solution through a column of an anion-exchange resin (De-Acidite FF or equivalent) in the OH^-, or better, the HCO_3^- form. If the resins are used in the reverse order to that just described the intermediate eluate is strongly alkaline, and thus neither procedure is satisfactory if, as in the case of sugars, the substances being desalted are sensitive to acid or alkali. This difficulty can be overcome by using a mixed-bed resin, made by mixing the two resins on the column, or obtained as Bio-Deminrolit (Zeo-Karb 225 + De-Acidite FF), or an equivalent. These mixtures are used in the same way as the individual resins; the effluent solution will contain only the neutral molecules, and will have a pH of about 7.

Mono-bed resins are regenerated with acid (HCl) for cation-exchangers and with carbonate-free alkali (NaOH) for anion-exchangers. In mixed-bed columns, back-washing with water causes the resins to separate as shown in Fig. 3.19. Acid is then passed in at A to regenerate the cation-exchanger, and subsequently alkali at B to regenerate the anion-exchanger. Both solutions emerge at C. The column is

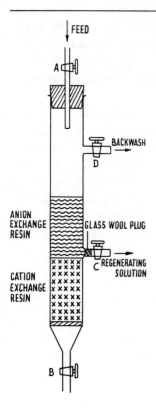

Fig. 3.19. Mixed-bed resin for desalting

washed with water, and air is blown in at B to remix the resins. An advantage of ion-exchange desalting is that it does not need complicated apparatus, and it avoids the chemical changes associated with the electrolytic method.

Electrodialysis

The two processes of electrolysis and ion-exchange can be combined by the use of ion-exchange membranes, which are ion-exchange resins made in the form of a flexible sheet, instead of in the usual form of beads. There are two types, cation- and anion-exchange membranes (for example, Permaplex C-20 and A-20 membranes, respectively). They are permeable to solvent and small neutral molecules, but not to larger organic molecules.

The cation-exchange type acts as an anion filter, allowing only cations to pass through it when a current is passed; an anion-exchange membrane similarly allows only anions to pass. The membranes are supplied in a moist state, in which they may be distinguished visually, the anion-exchanger being the paler in colour. If the membranes are allowed to become dry they contract considerably and become brittle, but soaking in water restores their flexibility; drying does not affect the ion-exchange properties.

A simple cell of the type shown in Fig. 3.20 can be

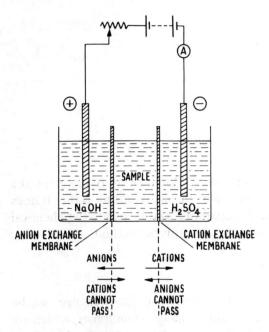

Fig. 3.20. Desalting cell for electrodialysis with ion-exchange membranes

used[31, 32]; it can be constructed easily from Perspex or similar material, although it is not easy to make the edges of the membrane watertight. Commercial versions of the apparatus are available. On a very small scale a high voltage can be used, but heating effects can be considerable, and for a sample volume as large as 2 cm³ it is necessary to use a low voltage. When a current is passed through the cell, the anions and cations migrate from the centre towards the appropriate electrode. Since migration from the electrode compartments is prevented by the membranes, the current falls rapidly as the concentration of ions in the centre cell is reduced. Inorganic ions, with their greater mobility, migrate first, but organic ions (including amino-acids) are lost by normal ion-exchange on the resin. Therefore the process is suitable only for desalting neutral compounds such as sugars. The solution in the centre cell remains substantially neutral during the desalting. The membranes can be used repeatedly, and they do not need regeneration, as that is taking place continuously during use.

Solvent extraction

It is not necesary to discuss extraction methods in detail. They have the advantage of causing the minimum of chemical change in the substances being desalted. With water-soluble substances such as amino-acids and sugars the method has its limitations. But if some of the substances present in low concentration in an aqueous solution are more soluble in an organic solvent, extraction with that solvent will desalt and concentrate them at the same time.

Application of the sample to the paper

The solution of the mixture to be separated is applied at a marked spot. It is usually allowed to spread to form a circular patch, although for separations on narrow strips or on circles it is often put on as a streak along a marked line. The part of the paper to which the drop is being applied should be kept fairly taut, and held horizontally, so that as the solution diffuses outwards it retains a circular shape, or remains as a compact streak. The zones do not exactly maintain their original shape as they migrate, but the better the shape at the start, the better will it be at the end. There should be nothing touching the underside of the paper while the spot is being applied. Thus the paper should not rest on the bench (nor, particularly, on any other sheet of paper to be used for chromatography). The best technique for loose sheets or strips is to clip one end to a suitable hanging rack and to hold the other end in the hand, while the sheet is drawn out to a horizontal position. Circles can be supported on a Petri dish, and square sheets for use in frames are spotted while they are in the frame.

The size of the spot depends on the scale of the experiment, but a diameter of about 0.5 cm should not be exceeded. This diameter is related to the thickness and the absorption characteristics of the paper, but generally the smaller the spot, the better the separation. The volume which can be applied to any particular paper to give a spot of this size is easily determined. The important quantity is not that volume, but the actual amount of the mixture which remains when the solvent has evaporated. If the solution is too dilute for the required amount to be applied at once, it can be 'concentrated' on the paper by applying a series of drops to the same place, allowing each to dry completely before applying the next. The spot should preferably be allowed to dry in air, but a gentle draught from a fan (or a domestic hair dryer, or an industrial glass-ware dryer) can be applied. It is not advisable to use hot air, particularly if the solution is acid, as it may cause blackening of the paper. If the mixture is being concentrated on the paper, or if many spots are being put on the same sheet, the drying time makes the procedure tedious. Simple devices have been designed to avoid this (for example, by Duncombe and Peaple[33]).

The temptation to apply too much of the mixture should be resisted. The minimum quantity of any component which should be present depends on the sensitivity of the locating reagent; the maximum is the amount which will migrate as a discrete zone. Excess of any component may mean that, because the rate of flow is fixed at a constant temperature by the paper/solvent system, this component cannot achieve equilibrium partition as it migrates, and thus it forms a streak which may obscure the position of other less abundant components. Preliminary experiments with unfamiliar mixtures should always start with conservative amounts; actual quantities in various cases are discussed on page 164.

There are several ways of applying the spots. One is the use of glass capillary tubes (melting point tubes are suitable, although a little wide), which should be of uniform diameter, and have reasonably square ends. Tubes of this sort can be roughly calibrated for semi-quantitative use, and, in any case, for a series of parallel separations it is easy to take the same volume each time. The end of the tube is held in contact with the paper until all the liquid has diffused out. The tubes should be regarded as expendable, and each should be used for only one solution.

There are advantages over the capillary tube method in the use of a platinum wire, fused in a glass handle, whose end is formed into a loop, as in the 'borax bead' test. The loop should be 2–3 mm in diameter, and should be slightly bent to form an angle with the wire, the free end being wound round. If the loop is dipped in the solution a drop is picked up which is easily transferred to the paper. Successive applications will result in drops of reasonably similar volume (about 4 mm^3) being put on each time. The loop is easily cleaned by rinsing and flaming (the latter may suffice to remove organic compounds, but the former should not be omitted when inorganic substances are being dealt with) and thus a series of comparison and test solutions can be accurately and quickly applied to the paper. Only when the test solution is in a very volatile solvent, such as ether or acetone, or when a streak is required rather than a round spot, will a glass tube be found to be better than a platinum loop. For accurate quantitative work a glass micro-pipette delivering, say, $10 \, mm^3$, or the more expensive but much more accurate Agla syringe, must be used. The former is sufficiently accurate for chromatographic separations, but the latter allows a smaller volume to be used.

Handling of sheets

Strict cleanliness is essential for paper chromatographic work, and dust, grease, and other adventitious substances should be avoided. If laminated plastics bench tops are not available, the ordinary laboratory bench should be covered with paper or plastics sheets. Marking must be done with an ordinary lead pencil (B or softer), since any form of ink will run in chromatographic solvents. The paper should be touched as little as possible on its surface with the fingers. Handling of small strips and circles presents no difficulty, as they can be held with forceps, or by the edges with one hand. Two-way square sheets in frames are similarly easy to deal with – in fact this is the main reason for the existence of this type of apparatus. It is in the handling of large sheets, up to 45 x 55 cm in size, that care is necessary. These sheets can be handled by a part which will not be reached by the solvent, but it is better to fasten them to a glass rod for carrying. It is convenient if the rods are of the correct length to fit into the tank. The sheet and rod can then be put into the tank together, and removed together. Ordinary 'spring' wooden clothes pegs form a cheap and simple way of fastening paper to hanging rods and racks.

Whereas handling of dry sheets presents little difficulty, and their insertion into the tank is no problem, much greater care is necessary when the wet sheets have to be removed. It is important that wet sheets should not touch each other, and that no solvent should be splashed on the sheet behind the solvent front. With the descending method, the recommended procedure is first to remove the residual solvent from the trough with a pipette, and then to clip the paper to the support rod at the ends (Fig. 3.21). The part of the paper dipping into the trough is cut off with scissors (taking great care, as the wet paper tears very easily), and rod and sheet are lifted out and transferred to a suitable drying rack of a size to support the rod at its ends. Racks corresponding in size to all the tanks in use in a particular laboratory should be constructed (Fig. 3.22). It is here that the frame apparatus again shows its convenience, as it is only necessary to lift the complete frame out of the tank. With other techniques the best procedure will be obvious in each case. The main consideration is that the sheets should be subjected to the

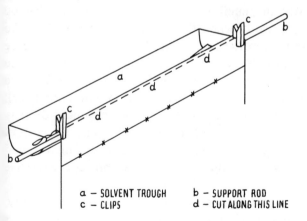

a − SOLVENT TROUGH   b − SUPPORT ROD
c − CLIPS   d − CUT ALONG THIS LINE

Fig. 3.21. Cutting large sheets before removal from the tank

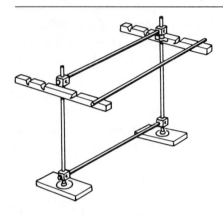

Fig. 3.22. Paper drying rack

minimum of handling while wet, and that they should be clipped to a suitable support as soon as possible, and left on that support during as many as possible of the subsequent operations.

The position of the solvent front should always be marked as soon as the sheet has been hung up to dry. It can be marked lightly in pencil at the edges of the sheet, and be drawn in more firmly when the sheet is dry. Preferably, the paper should be allowed to dry naturally in the air, in a place where it cannot be affected by fumes. Drying can be accelerated by means of a draught from a fan. This may be warm (for example, a low setting on a domestic fan heater), but it is not advisable to use hot air, which might destroy some of the constituents of the mixture. The air current should pass parallel to the surface of the sheets, which should be weighted with a 'bull-dog' clip to prevent excessive flapping.

Most chromatographic solvents evaporate quite quickly leaving no residues. One exception is phenol, which needs about four hours in a strong draught for complete evaporation. It is always desirable to achieve complete removal of the solvent, to avoid any adverse effect on the locating reagent but it is particularly important in two-way chromatography that all traces of the first solvent should be removed before the second run. For this reason it is best to use the more volatile solvent first (thus, butan-1-ol/acetic acid/water before phenol/water). But no matter how the first solvent is removed, there are always differences in the final results if the order is changed. For reproducible results the same order must always be used.

A number of drying ovens are available which are suitable for chromatography, some being specially designed for this purpose. Mostly they have removable racks and electric heatirg elements, and employ a forced draught. Such ovens are most useful when the detecting reagent requires the application of heat, but at low temperatures they can be used for drying. An important advantage of a closed drying cabinet is that it gives greater freedom from airborne contamination. Sheets dried in the open laboratory, or in fume cupboards, should be protected from dust as much as possible. If an ordinary oven is used for drying, the sheets should be arranged to hang vertically, and precautions should be taken to prevent an accumulation of inflammable solvent vapours.

Location of separated substances

The success of a chromatographic separation depends ultimately on the location process. Coloured substances are, of course, visible as separate patches at the end of the run. Colourless ones require chemical or physical detection. It should be a routine that chromatograms are examined under ultra-violet light before and after the application of any other method. The wave-lengths usually recommended are 370 nm and 254 nm. Some of the substances present may appear as fluorescent patches, although there are cases where substances appear as dark patches on a fluorescent background. That happens more often after a reagent has been applied.

Two other physical methods, applicable only to radioactive substances, are auto-radiography and counting (p. 165). They can be used to supplement chemical methods of detection and identification of the active material.

Chemical methods of detection are the most important. The reagents used are commonly referred to as 'locating reagents'. This section deals with the location of separate zones into which the various substances have been moved, and the locating reagent must reveal as many of these areas as possible. Identification of the substances is a separate problem (p. 153), although obviously with the right choice of reagent location and identification can be achieved at the same time.

The locating reagent is applied in some suitable solvent, which is allowed to evaporate. Ideally the solvent should be one in which all the separated substances are insoluble, but that is not always possible, as the other requirements for the solvent may conflict with it. The classical way of applying the reagent is by spraying, but dipping has many advantages[34]. Spraying is usually done with some simple type of glass unit. A number of ways of making them have been published, and several patterns are obtainable commercially. They are usually supplied with a rubber bulb type of hand blower, but it is not easy to obtain good results in that way, and an electric compressor, or a compressed air supply, is much more satisfactory. Some units provide an even spray by incorporation of a reservoir of a volatile propellant such as dichlorodifluoromethane attached to a valve and spraying nozzle, with an interchangeable container for the reagent solution. It is essential to cover the sheet evenly with the solution without so saturating it that the solution flows over the surface, because that can cause disturbance of the zones. The spray is therefore held about two feet from the sheet, and moved slowly from side to side and up and down; illumination from behind helps in seeing whether any unsprayed areas remain. There are several disadvantages in the spraying method:

The quality of the spray units varies. Some are excellent, giving a good fine mist. Others tend to throw out large drops, or to give a very uneven spray. These are common faults and are not confined to any one design or manufacture. The glass units are expensive and somewhat fragile. Small metal or plastics scent sprays are not recommended.

However carefully the spray is used there is a considerable waste of solvent and reagent. The solvents are cheap, but some of the reagents are quite expensive.

There is some fire and toxicity hazard.

The solvent used for spraying must not be too volatile; otherwise there will be extensive evaporation from the small drops, and hence poor coverage. On the other hand, rapid evaporation from the paper is required to avoid diffusion of the separated spots. The solvents most commonly used are ethanol, propanol, butan-1-ol, or chloroform, or mixtures of them. Aqueous mixtures can be used, but too much water should be avoided if possible, because of its weakening effect on the paper. Spraying should always be done in a fume cupboard with a good draught; lining with disposable paper or plastics sheets prevents the spray from soiling the inside of the cupboard. The spray should always be cleaned immediately after use, to prevent blocking of the jet.

In the dipping technique the reagent solution is put in a suitable shallow dish, and the sheet or strip is drawn through it in a single 'see-saw' movement. Any shallow dish of the appropriate size can be used, but special polythene or polythene-coated metal trays can be bought. The technique is simple for small sheets, strips, and circles, but more practice is required with larger sheets, although, since the extremities may not need to be immersed, there should be no great difficulty. There are several advantages in dipping:

The only apparatus is the dish, which is cheap and robust. Even coverage of the sheet is automatically assured. A more volatile solvent can be used, thereby reducing the time needed for drying, and hence reducing the danger of diffusion of the spots during drying. Acetone is the preferred solvent, but there is no reason for not using the same solvents as for spray reagents.

Unfortunately it frequently happens that the separated substances are very soluble in the best solvent for the locating reagent. In such cases dipping would cause unacceptable diffusion or loss of material, and spraying is essential.

After application of the initial reagent, some further treatment is often necessary. In the case of amino-acids the first reagent applied is normally ninhydrin (indanetrione hydrate). With amino-acids it does not give colours until about twenty-four hours later at room temperature. At 105°C colours appear after about four minutes, although with some background colour and some colours produced by compounds which are not amino-acids. Many inorganic separations are done in the presence of acids, and after application of the reagent the sheet must be rendered alkaline. This is done most easily by holding the whole sheet with forceps in a large tank at the bottom of which is some 0.880 ammonia.

A technique developed by Smith[34] is that of multiple dipping. The sheet is treated with one reagent, the zones are marked, and the sheet is then treated with a second reagent. This may bring up new coloured spots, or may modify an existing spot in a characteristic way. For example, hydroxy-proline with isatin gives a pale blue spot, which changes to a dark purple when treated with Ehrlich reagent. This method of multiple dipping should not be confused with procedures where a sheet must be treated with several solutions in succession to bring up the initial spot, which is not further modified.

Sometimes it is not easy to tell whether a mark on the paper is a definite spot or not. Aids for deciding, apart from the routine examination in ultra-violet light, are to examine by transmitted light, or to look at the 'back' of the sheet (that is, the side opposite to that on which the sample was applied), where there may be less background colour and spots may show up better.

The selection of a locating reagent is purely empirical; the reagents are those which give a colour reaction with as many of the substances present as possible. Some of the reagents used do not give a colour reaction in the usual sense; for instance, a good reagent for reducing sugars is silver nitrate, which gives dark grey spots. In the Appendix (p. 360) there will be found a list of reagents suitable for various separations, and also some notes on the colour reactions of amino-acids.

Identification of substances

A theoretical treatment of partition chromatography, in which the column or sheet is considered to be analogous to a distillation column, was worked out by Martin and Synge[35], and was described by Brimley and Barrett[36] (see also p. 260). In that treatment the R_f value, which is defined as

$$R_f = \frac{\text{distance moved by substance}}{\text{distance moved by solvent front}}$$

was shown to be related to the partition coefficient of the substance between water and the mobile phase. At one time it was thought that the R_f value might become an important physical constant for the substance, but experience has shown that it is of rather less value than had originally appeared. That is hardly surprising in view of the number of factors which determine the R_f value in various circumstances:

The solvent. Because of the importance of the partition coefficient, quite minor changes in solvent composition can cause considerable changes in R_f values, and in the relative R_f values of two substances, even to the extent of reversing their positions on the chromatogram.

The temperature. The concentrations in the two phases at equilibrium depend on the partition coefficient. How nearly these concentrations approach the equilibrium values depends on the rate of flow of the mobile phase, which in turn depends partly on the viscosity of the liquid. Since both viscosity and partition coefficient are temperature-dependent the temperature has an important effect on R_f values. On fast papers a rise in temperature improves the efficiency of separation considerably, provided that the loading is low. The effect on slow papers is very much less significant.

The size of the tank. The effect of tank size has been discussed under the heading of equilibrium. The volume of the tank affects the homogeneity of the atmosphere, and thus the rate of evaporation of solvent components from the paper. If a large tank is used there is a tendency to allow a longer run; as the solvent composition changes along the sheet, the partition coefficient will change also. Attention has been drawn[50] in this connection to the desirability of always having the same distance between the start line and the position of solvent feed. These two factors of evaporation and composition influence the R_f value.

The paper. For a given mobile phase at a fixed temperature the flowrate is determined by the nature of the paper. This is probably the main way that the paper influences R_f values. Differences in R_f values which are observed in upward, downward, and horizontal flow may be ascribed to this cause. The incursion of adsorption and ion-exchange effects also varies from one paper to another.

The nature of the mixture. The various substances are undergoing partition between the same volumes of stationary and mobile phase. They are nearly always likely to influence each other's solubility characteristics, and, thus, their R_f values.

When lists of R_f values are published it is necessary to state all the conditions used in their measurement, and also the volume of solvent taken. It has been shown that that volume can affect R_f values, although by what mechanism is not clear. To measure R_f it is necessary to locate the solvent front. For some separations, notably of sugars, the R_f values are rather close together, and to give a good separation a very long strip would be needed. Since the R_f values concerned are rather small, the solvent is allowed to run off the sheet, and thus the values cannot be determined. In such a case use is made of an R_x value, based on the fact that under any given conditions the relative R_f values remain the same. Therefore any convenient substance can be used as a reference, the migrations of the others being referred to it; thus

$$R_x = \frac{\text{distance moved by substance}}{\text{distance moved by } x}$$

For sugars, glucose is used as the reference substance (R_g values).

R_f values are usually expressed as a fraction, to two decimal places, but it was suggested by Smith[11] that a percentage figure should be used instead; that is, 0.72 or 72. The fraction system is the more widely used. The difference in R_f values required for two substances to be separated depends on the size of the spots and the length of the solvent flow. Thus, if the initial spot is 1.0 cm in diameter, and if that size is maintained during the run, two substances forming spots with centres 1.0 cm apart will just be separated. If the solvent flows 30 cm this gives an R_f difference of 0.03. In practice, spreading of the zones occurs, and a longer run would be needed (or a smaller initial spot).

The simplest way of measuring R_f is with an 'elastic ruler', which can be made from a piece of elastic material marked in 10 equal divisions from 0 to 10. The zero end is placed at the origin, and the 10 end is stretched to the solvent front, the R_f being read off directly at the point where the ruler passes over the spot. Other more elaborate devices have been described[37]. If many runs are done under the same conditions, using the same length of flow every time, a piece of tracing paper with appropriately spaced lines drawn on it can be put over the sheet, and the R_f can be read directly. In determining R_f it is necessary to measure to the centre of the zone. If the spot is irregular in shape, or if the substance has 'streaked', it is not always possible to locate the centre with any accuracy, but with a reasonably compact but irregular spot an attempt should be made to locate a 'centre of gravity' for measurement. No useful conclusion about R_f can be made in cases of bad streaking; the separation should be repeated with some other solvent.

Identification of substances by comparison of experimental with published R_f values is not always very certain; R_x values are better, but discrepancies will still be found. If many similar samples, such as biological or pathological specimens, are being regularly examined, one comes after a short time to recognise a normal pattern. The absence of any component, or the presence of any unusual one, are both immediately recognised. Actual identification of an unknown substance can be done by repeating the separation. The area of paper containing the unidentified substance is cut out (located by comparison with the first sheet) and the substance is eluted with a suitable solvent. The solution can be examined by chemical methods, or it can be chromatographed in a mixture with a series of separate reference substances. When only one spot is obtained in more than one solvent with the same reference substance, the two can probably be taken as identical. Some confirmation can be obtained by mixing the reference substance with all the other substances in the original mixture, and comparing a chromatogram of that mixture with the first one. Reference mixtures can be made up on the paper, using the method described for 'concentrating' the original solution. Evidence about the identity of new compounds obtained by chromatography should always be supplemented by chemical evidence.

Another method of identifying substances is by means of characteristic colour reactions. For inorganic separations it is probably the most useful, but it is of less value for organic separations where the constituents of a mixture are chemically more similar. The technique of multiple dipping overcomes the deficiency to some extent; it is complementary to the procedure of running several identical chromatograms on one sheet, cutting into strips when dry, and treating each piece with a different reagent.

Reversed phase methods

If the substances being chromatographed are only very sparingly soluble in water, they merely move with the solvent front, and thus no separation results. In such a case it may be advantageous to impregnate the paper with a non-aqueous medium, to act as the stationary phase. In 'normal' chromatography the stationary phase, being aqueous, is more polar than the mobile. The technique where the mobile phase is the more polar is called 'reversed phase' chromatography. The mobile phase is not necessarily water, although the mixtures used normally contain some water.

A number of substances have been employed as supports

for the stationary phase, among them rubber latex, olive oil, silicone oils, and similar liquids. The greatest difficulty is to get even impregnation of the paper. The usual procedure is to draw the sheets with a see-saw motion through a solution of the substance, and then to allow the solvent, which should be a fairly volatile one, to evaporate. Some papers, for example Whatman No. ST81, impregnated with silicones in manufacture are now available commercially. Their use removes the difficulty about even impregnation in cases where a silicone support is acceptable.

If the paper is treated with rubber latex or silicones it is rendered water-repellent, and it absorbs the organic component of the solvent mixture, which then becomes the stationary phase, in preference to the water. Where liquids such as olive oil are used they perform a similar function, but may also act as the stationary phase themselves. There must also be some modification of the adsorption and ion-exchange effects of the paper.

Reversed phase methods can be applied to all the various forms of paper chromatography. The general procedure is the same as for normal methods. The solvent for the locating reagent must be one in which the stationary phase support is not soluble.

Ion-exchange papers

A combination of the specificity of ion-exchange with the convenience of paper chromatography is afforded by ion-exchange papers. There are two kinds. One consists of cellulose where acidic or basic groups have been introduced by chemical modification of the —OH groups; the other is made by blending an ion-exchange resin with cellulose, and making sheets with the mixture in the normal way. Paper so obtained contains about 45% of resin by weight. For reasons connected with patents the resin-loaded type of paper is not so widely available as the modified cellulose type. Examples of the latter are given in Table 3.3. In addition to paper sheets, the modified celluloses are supplied as powders and flocs for use in columns, the techniques being the same as those used for cellulose powder partition chromatography and ion-exchange resin chromatography. Special powder forms are made for spreading on thin-layer plates (p. 284). Knight[47] made a study of the properties of the two types of paper.

Examples of the uses of ion-exchange papers include the separation of amino-acids[11, 47], metals[11], and other ionic species. The apparatus used is that of ordinary paper chromatography, and the methods are in general the same; for instance, similar factors influence the choice between ascending and descending techniques. The test solution can be applied to the dry paper in the usual way, but there is no need, within reason, to restrict the volume applied, nor to let the spot dry, as the ions in solution will be held at definite points on the paper by ion-exchange. In many cases it is necessary to use some other 'form' of the paper than that in which it is supplied. Thus carboxymethyl cellulose paper is supplied in the sodium form, and for the separation of some mixtures of metals it is best to convert it to the magnesium form[11]. The conversion is done in a descending chromatography tank, by allowing a solution of magnesium chloride to flow down the paper overnight. The test solution can then be applied to the wet sheet, and the chromatographic run is begun immediately, elution being with the same solution of magnesium chloride. The wet method is applicable only to descending chromatography. It is better than the dry method when pretreatment of the paper is necessary. The only disadvantage is the absence of a solvent front, which means that the run must be on a time basis, and R_f values cannot be measured.

When the ion-exchange properties of the paper are being

Table 3.3 *Whatman modified cellulose ion-exchange materials*

No.	Type	Name	Ion-exchange group	Flowrate water (upward) mm/30 min	Ion-exchange capacity mmol H$^+$ cm^{-2}
P81	strongly acid	cellulose phosphate	$-O \cdot PO_3 H_2$	125	18.0
CM82	weakly acid	carboxymethyl cellulose	$-CO_2 H$	110	2.5
DE81	strongly basic	diethylaminoethyl cellulose	$-C_2 H_4 \cdot NEt_2$	95	3.5
ET81	weakly basic	Ecteola cellulose	tert. amino—	125	2.0

used, the eluting solution must contain some ion which can displace those to be separated. Choice is made in the same way as in ion-exchange resin chromatography on columns. One-way separations of amino-acids and of metals have been reported on both anion- and cation-exchange paper with aqueous buffers. pH is important, particularly in the case of the weakly acidic and basic papers. The ion-exchange papers still retain the cellulose matrix, however, and if the exchange properties can be suppressed, the sheet can be used to give the same sort of separation as does ordinary paper. Two-way separations with different buffers in each direction do not give useful results, but the combination of ion-exchange separation in one direction with a conventional chromatographic separation in the other seems to have considerable potentialities. As an illustration of the technique, the following conditions were reported by Knight[38] for the separation of a number of amino-acids:

(*a*) cellulose phosphate paper — H^+ form.
solvent 1: 0.02 mol dm^{-3} sodium buffer, pH 4.7
solvent 2: *m*-cresol/1% ammonia.

(*b*) DEAE cellulose paper — free base form.
solvent 1: 0.02 mol dm^{-3} acetate buffer, pH 7.5
solvent 2: *m*-cresol/1% ammonia.

The locating reagent in each case was ninhydrin, which is not entirely satisfactory for use on some anion-exchange papers, where the paper itself is a weak ninhydrin reactor and gives a high background colour.

In the first solvent for each separation the amino-acids were in the cationic form in the pH 4.7 buffer on cellulose phosphate paper and the anionic form in the pH 7.5 buffer on DEAE cellulose. In the second solvent they were in the anionic form in both cases, and thus on cellulose phosphate paper no ion-exchange occurred, the compounds behaved as they do on ordinary paper. On the DEAE cellulose the partition solvent probably inhibits the ionisation of the exchange groups, giving a separation mainly of the ordinary chromatographic type. The separation obtained is illustrated in Fig. 3.23, with a diagram of the same separation carried out by a normal two-way procedure for comparison. It will be observed that the resolution is much better on the strongly acidic cation-exchange paper, and that the acids are more evenly spread over the sheet.

Knight[39] also described an even better separation, on paper impregnated with Zeo-Karb 225 (W.R. 1.5–2.0). In this case none of the separated acids appeared on the diagonal of the paper through the origin, and some acids were separated which are usually difficult to deal with, such as leucine and isoleucine.

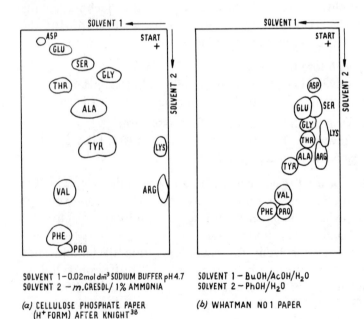

SOLVENT 1 – 0.02 mol dm³ SODIUM BUFFER pH 4.7
SOLVENT 2 – *m*.CRESOL/ 1% AMMONIA

(a) CELLULOSE PHOSPHATE PAPER
(H⁺ FORM) AFTER KNIGHT³⁸

SOLVENT 1 – BuOH/AcOH/H₂O
SOLVENT 2 – PhOH/H₂O

(b) WHATMAN NO 1 PAPER

Fig. 3.23. Separation of amino-acids

The advantages of the ion-exchange paper method appear to be increased resolution, which makes the use of smaller sheets convenient, and a saving in time. A two-way separation may be done in one day, or in one day and one overnight run, as against the usual two overnight runs of conventional paper chromatography.

Quantitative methods

Chromatography is a method of separating substances in a mixture. Quantitative use of the technique requires not only a quantitative separation, but also quantitative location and evaluation of the substances present. If a good chromatographic separation can be obtained, then the quantitative application depends solely on the last factor. A satisfactory qualitative separation is not necessarily useful quantitatively. The quantitative finish can be either by estimation of the amount of substance in the spot on the paper, or by removal of the substance from the paper, and analysis of the separate fractions by conventional quantitative techniques.

It cannot be emphasised too strongly that rigid standardisation of procedure is essential if good quantitative results are to be obtained. The original spot is applied with a calibrated capillary tube, a micro-pipette or an Agla syringe; 10 mm^3 is usually a convenient volume. Drying of the spot must be done under standard conditions of time and temperature. The solvent must be made up with extreme care as to proportions, equilibration must be in a standard manner, the length of run must be the same each time, the temperature must remain constant throughout, and there must be a standard time and temperature for drying the sheet. The locating reagent, if the coloured spot is being used for the measurement, must be applied in an exactly reproducible way, and any after-treatment, such as drying or exposure to ammonia, must be for a standard time.

The amount of substance which should be put on the paper for a chromatographic separation varies considerably. For instance, the minimum amount for some amino-acids is about 0.1 μg, but for others it is about 20 μg. For sugars 30–40 μg may be needed. The quantities have to be somewhat larger for two-way separations. Separation of metals can be done with as little as 0.1 μg, in some cases, but larger amounts, up to about 200 μg, may be needed in others, particularly if the substances are to be eluted.

Evaluation of the substances on the paper

Visual comparison of spots. A number of chromatograms are run on the same sheet, the reference solutions containing known amounts of the substance present. Several reference solutions of different concentrations must be used, and it is important that each shall contain all the constituents of the test solution. This method depends on the meticulous standardisation already mentioned, but, given that, it is surprisingly accurate, particularly in the estimation of metals. The use of CRL/1 papers is recommended, and on this scale quantities of 0.2–2.0 μg can be estimated to about 3–5%; for organic separations colour matching is usually only possible to within about 10%.

Physical measurement of coloured spots. Spectrophotometers can be used or adapted to measure the amount of substance either by reflectance or transmission. Densitometers of various types are specially made for the easy scanning of chromatographic strips. They find their most important use in the examination of strips after zone electrophoresis.

Measurement of the area of the spot. The area of the spot is proportional to the logarithm of the concentration of the substance in the original solution. Measurement of the area is made difficult by the lack of a sharp boundary, but this method of estimation has been used in a few cases.

Radioactive measurements. The simplest way of detecting radioactive substances is to scan the strip with an end-window Geiger-Müller tube, and to mark the positions of the spots. The area of paper containing each spot can then be cut out, put on an ordinary planchette and counted accurately in a lead castle. A better method is to use an automatic scanning device, which can be calibrated to give a quantitative evaluation of the spots. The result can be confirmed by elution of the substances and measurement of the active material in the eluate. For two-way chromatograms auto-radiography is easier and more efficient than counting for detecting the positions of the substances. The sheet is placed in contact with a sheet of X-ray film, which, after a certain exposure and development, is found to have been blackened in the areas corresponding to the active spots. The autoradio-graph can be used as a guide for cutting out the appropriate areas from the chromatogram for elution and counting.

Removal of the substance from the paper

The methods so far described (except counting and auto-radiography) require the application of a chemical locating reagent to show the position of the spot, and the coloured patch forms the basis of the estimation. The complex can sometimes be eluted and estimated colorimetrically, but if the chemical change involved is not acceptable the unaltered material must be eluted. It is then necessary to make two chromatograms under identical conditions, to apply the locating reagent only to one, and to use that to mark the position of the substance on the other, which is then eluted.

The technique is really that used on the CRL/1 papers, but on a larger scale. The method of Pollard[40] is to use a strip cut longitudinally into two or three (Fig. 3.24 (*a*)). The same volume of test solution is applied to each origin, in the form of a narrow line. After development the strips are separated;

the locating reagent is applied to one, which is used as a guide in cutting the other for elution of each piece separately. For larger quantities a sheet, with a tracer strip at each side, can be used. In that case the tracer spots will contain only a fraction of the amount in the main part of the paper, but the method works quite well (Fig. 3.24 (*b*)).

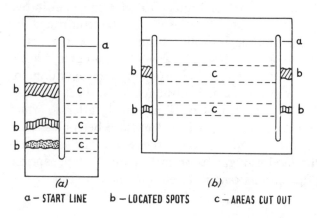

a – START LINE   b – LOCATED SPOTS   c – AREAS CUT OUT

Fig. 3.24. Method of using sheets in quantitative work

The elution can be done by immersing the piece of paper in a solvent, by extraction in a Soxhlet apparatus, or by using some special arrangement giving in effect a downward chromatographic flow through the paper. For inorganic separations the pieces of paper can be ashed, and the residue can be taken up in acid. This method does not give as good results as elution. The solutions thus obtained can be analysed by any conventional method; the ones most used to follow chromatography are colorimetry and polarography.

It is not absolutely necessary to find a chromatographic method which will separate quantitatively all the components of a mixture. For instance, if a metal A is to be estimated by polarography, and the only metal present in the mixture which will interfere is B, it is only necessary to separate the mixture into two groups, one of which contains all the A but no B; the fact that A is still mixed with other metals is now immaterial, as long as its position can be found. In other cases it may be possible to arrange that only the wanted substance moves from the origin, or it is the only one which does not move.

Quantitative estimation of organic substances is rather more difficult than that of metals, because there are fewer available methods for examination of an eluate. Organic estimations are usually done on the paper, and thus require that each substance shall be quantitatively separated from all the others.

Faults in paper chromatograms

Ideally, the final spot which is revealed on application of the locating reagent should be the same shape, and as compact, as the original spot. Some spreading always occurs, and it has to be accepted; the final spot is often heart-shaped, and if it is compact that is acceptable also. Some more serious faults, and their causes, are described below. Some of them are not really faults, as they are inevitable with the conditions employed, but their interpretation is necessary if the chromatogram is to be useful. The method of rectifying genuine faults will be obvious.

Multiple spots[41]

Occasionally, when a substance thought to be a single entity is chromatographed, two spots are obtained; they may be quite distinct, or one may be a 'ghost' spot, having the same shape as, and being close to, the main one. There can be various explanations if the substance is inorganic. In chromatography of metal ions, if the solvent contains a different anion from that in the original solution, there may be competition between the anions for the metal ions, with the result that two spots are obtained, one for each of the two salts of the metal. The metal may be present in two oxidation states, such as Fe^{2+} and Fe^{3+}, which migrate at different rates in the same solvent, or there may be two ions, such as Bi^{3+} and BiO^+. Finally, the metal ion may form two different complexes with the solvent.

In organic separations, the substance may be present in two different forms. For instance, an amino-acid may be present as its cation and as the dipolar ion. Chromatography of a substance such as lactone will always give a spot due to the free acid, because water is always present. Sometimes substances are contaminated by oxidation products. For example, specimens of the amino-acid cysteine often contain traces of cystine. In other cases traces of position isomers or stereoisomers may appear as separate spots (optical enantiomers normally move as a single spot, although some partial resolutions have been reported).

Tailing

If too much of the mixture is put on, or the solvent flow is too fast, the substance cannot attain the partition equilibrium required to give a discrete spot. It then tends to be spread out over a large area of the paper, and to be left behind by the advancing solvent. Tailing may also be caused by adsorption effects.

Edge effects

Spots very near the edge of a strip may tend to spread along the edge; the diffusion may be due to a local higher concentration of mobile phase in that area, or it may be due to a higher local rate of evaporation of solvent from the edge, giving abnormal partition effects.

Applications of paper chromatography

The literature on analytical methods and on the investigation of natural compounds shows that there can hardly be any field in which paper chromatography has not found some use, although it is still most widely employed in separations of a biochemical nature. It is much used as a research tool, and is also used extensively for routine analyses, particularly for new separations where no classical method exists. Some of the principal uses, with examples, are summarised below. It is emphasised that the actual separations mentioned are given merely for illustration, and it is not suggested that the list is in any way comprehensive.

Clinical and biochemical. Separation of amino-acids and peptides in the investigation of protein structures. Routine examination of urine and other body-fluids for amino-acids and sugars (this is most important, as it can be used for diagnosis of a number of pathological conditions, with the 'standard map' technique). Separation of purine bases and nucleotides in the examination of nucleic acids. Separation of steroids.

General analytical. Analysis of polymers[42]. Detection and estimation of metals in soils and geological specimens[8]. Investigation of phenolic materials in plant extracts[43]. Separation of alkaloids. Separation of radio-isotopically labelled compounds.

It is perhaps only fair to mention fields in which paper chromatography has not been very successful, and where other forms of chromatography have been found to be better. One is the separation of volatile unreactive substances such as hydrocarbons. Another is the separation of the more volatile fatty acids. Quantitative and preparative separations are mostly more efficient when done by other forms of chromatography.

Although published work in which paper chromatography

is used continues to appear in considerable volume, use of the method seems to have passed its peak. Developments in thin-layer chromatography and gas chromatography, and the renaissance of column methods (p. 18), with the increased speed and degree of instrumentation that they offer, have diverted attention from the cheaper and less sophisticated paper methods. Paper chromatography is a technique in which the individual worker must devise the best solution for his own problems, because there are so many variables and no 'standard' procedure of universal application. It is therefore not surprising that it should be replaced by methods which lend themselves to automation and instrumental operation more readily, even though some large-scale users have been able to automate certain parts of the paper chromatography process[48].

Zone electrophoresis

Separation of charged particles by differential migration in an electric field is an old-established technique, which, in the form of the moving boundary method of Tiselius, is not obviously related to chromatography. After the introduction of paper chromatography, attempts were made to improve separations of ionic substances by the application of an electric field[44], and this led to the development of 'paper electrophoresis', the principle of which is illustrated in Figs. 3.25 and 3.26.

A paper strip, suitably supported, dips at its ends into electrode vessels containing a buffer solution, acting as the electrolyte. The paper is soaked in the buffer, and the sample is applied at some point on the strip as a thin transverse streak (by the same method as used in paper chromatography). The electrodes are connected to a d.c. source, and current is allowed to flow for a predetermined time. The strip is removed from the apparatus, dried, and treated with a

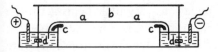

a – PAPER STRIP   b – POSITION WHERE
                           SAMPLE APPLIED
c – PAPER SUPPORT d – BUFFER

*Fig. 3.25. Apparatus for low-voltage
paper electrophoresis*

locating reagent – again as in paper chromatography. The separated substances then appear as a series of bands, whose distance from the origin depends on the charge on the ion, its mobility in the field applied (and thus on the applied voltage and the current), and the pH of the buffer (Fig. 3.26).

Whereas in chromatography the separation is achieved by means of a flowing solvent which differentially moves the various solutes, in electrophoresis the electrolyte is stationary and the flow of ions occurs by virtue of their charge. The only movement of solvent is the electro-osmotic flow (produced by ionic charges induced on the supporting medium), which is only slight, but which may restrain the movement of some of the ions of lowest mobility.

This technique, with its similarity to paper chromatography, is called 'paper electrophoresis'. The paper acts as a support for the electrolyte, and also for the separated substances. It restrains their diffusion in the buffer solution, and holds them on drying so that the locating reagent can be

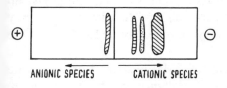

ANIONIC SPECIES   CATIONIC SPECIES

Fig. 3.26. Electrochromatogram

applied. The nature of the paper has very little bearing on the separation, and, since other supports are now available the terms 'zone electrophoresis' or 'electrochromatography' are commonly used. Electrophoresis can be used in conjunction with conventional chromatography. A paper sheet cut as shown in Fig. 3.27 can be used, in suitable apparatus, to give useful two-way separations of substances such as amino-acids and similar polar species.

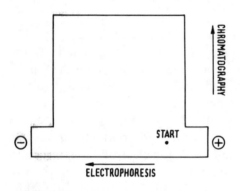

Fig. 3.27. Arrangement of paper sheet for combined electrophoresis and chromatography

The most widely applicable methods are those known collectively as 'low-voltage electrophoresis', in which the applied voltage is up to 1000 V. Voltages from 1 kV to 10 kV can be used in special apparatus for a rather restricted range of separations (high-voltage electrophoresis).

Several different supports and numerous designs of apparatus can be used for electrophoresis. The supports can be classified broadly as strips, gels, and thin layers; they may be totally inert or they may have a physical effect on the separation.

Paper. Whatman No. 1 and No. 3MM in strips 3 or 5 cm wide are most often used. 3MM has a slightly greater wet strength, which is an advantage when aqueous solutions are being used.

Since the paper is thicker it will pass a higher current than No. 1 for a given voltage. Apart from the use of still thicker papers for preparative separations there is nothing to be gained by changes in the type of paper, except that ion-exchange papers can be used to improve separations of some mixtures by the two-way electrophoresis/chromatography technique[45].

Cellulose acetate. For many purposes strips of cellulose acetate are preferred, since they give sharper bands, and are more easily rendered transparent for photo-electric scanning. Their chemical composition is about that of cellulose diacetate, and some are made in a porous form which improves the separation. They are available in a range of sizes and thicknesses.

Gels. Gels made from starch, agar, or polyacrylamide can be used as buffer supports. The gel is used in the form of a block, requiring a special carrier to hold it in the electrophoresis chamber. Starch and polyamides exert a slight molecular sieve effect, and therefore have a better resolving power in electrophoresis than most other support media.

Thin layers. Various groups of polar compounds, such as amino-acids, phenolic compounds, or amines can be separated by thin-layer electrophoresis or combined two-way electrophoresis and thin-layer chromatography. Glass plates, spread with silica gel or alumina (p. 286) are put in the apparatus in the position usually occupied by the paper, with paper strips overlapping the ends and dipping into the electrode compartments.

Apparatus. The general principle of the design of low-voltage electrophoresis apparatus is shown in Fig. 3.25. The essentials are two compartments to hold the buffer and electrodes and a suitable carrier for the support medium, such that its ends are in contact with the buffer compartments. The design of the carrier depends on the medium (strip, gel block, thin-layer plate) and Fig. 3.25 illustrates the use of paper strips. It will be noted that the paper does not dip into the electrode compartments, but into separate compartments connected by wicks with the anode and cathode cells. The purpose is to restrain diffusion of buffer electrolysis products along the paper, and to maintain the pH at the ends of the strip. In more recent designs a labyrinth construction replaces the wicks. The apparatus is enclosed to avoid evaporation from the paper, and sometimes provision is made for external cooling. The strip is not supported throughout its length, but is stretched as tautly as possible across the end supports. A power pack supplying up to 500 V, or even 1000 V, is needed. It can be of the constant current or constant voltage type, but the former gives better results.

Low-voltage electrophoresis can be used in principle to separate any ionic substances. In practice its main application is in the examination of biological and clinical specimens for amino-acids and proteins. The latter, in particular, are more easily separated by electrophoresis than by chromatography. There are numerous research and routine examinations of serum, plasma, and other similar specimens which are done in this way. The preferred support media for these separations are cellulose acetate strips and one of the various forms of gel. The proteins are located by staining with a dye, and they can be estimated with fair accuracy with an automatic or manual scanner. Sugars can be separated in a borate buffer, in which they form complex ions. They are located with the usual chromatographic reagents.

High voltage electrophoresis is a rather specialised technique, requiring special experimental precautions and conditions, and is thus of less general use. Large charged molecules, such as proteins, which are readily and conveniently separated by low-voltage electrophoresis, cannot be separated with high voltages. The technique works best with small ions, particularly with amino-acids.

REFERENCES

1. Consden, R., Gordon, A. H., Martin, A. J. P., *Biochem. J.*, 1944, **38**, 224.
2. Burstall, F. H., Davies, G. R., Linstead, R. P., Wells, R. A., *J. Chem. Soc.*, 1950, 516.
3. Pollard, F. H., McOmie, J. F. W., Elbeih, I. I. M., *J. Chem. Soc.*, 1951, 446.
4. Pollard, F. H., McOmie, J. F. W., *Chromatographic Methods of Inorganic Analysis*, Butterworth, London, 1953.
5. Hanes, C. S., *Canad. J. Biochem. Physiol.*, 1961, **39**, 119.
6. Wade, E. M., Matheson, A. T., Hanes, C. S., *ibid.*, 1961, **39**, 141.
7. Hanes, C. S., Harris, C. K., Moscarello, M. A., Tigane, E., *ibid.*, 1961, **39**, 163.
8. Hunt, E. C., North, A. A., Wells, R. A., *Analyst*, 1955, **80**, 172.
9. Pollard, F. H., Banister, A. J., *Analyt. Chim. Acta*, 1956, **14**, 70.
10. Datta, S., Dent, C. E., Harris, H., *Science*, 1950, **112**, 621.
11. Smith, I., *Chromatographic and Electrophoretic Techniques*, Heinemann, London, **1**, 1960.
12. Ridgway Watt, P., Green, J., *Chem. and Ind.*, 1959, 1543.
13. Rutter, L., *Nature*, 1948, **161**, 435.
14. Proom, H., Woiwod, A. J., *Chem. and Ind.*, 1953, 311.
15. Kawerau, E., *Chromatographic Methods*, 1956, **1**, No. 2.
16. McDonald, H. J., McKendall, L. V., Bermes, E. W., *J. Chromatog.*, 1958, **1**, 259.
17. Anderson, J. M., *ibid.*, 1960, **4**, 93.
18. Tata, J. R., Hemmings, A. W., *ibid.*, 1960, **3**, 225.
19. Cassidy, H. G., *Analyt. Chem.*, 1952, **24**, 1415.
20. Clayton, R. A., *ibid.*, 1956, **28**, 904.
21. Martin, E. C., *J. Chem. Soc.*, 1961, 3935.
22. Cartlidge, J., Tipper, C. F. H., *Chem. and Ind.*, 1959, 852.
23. Schwarz, V., *ibid.*, 1953, 102.
24. *idem.*, *Biochem. J.*, 1953, **53**, 148.
25. Linstead, R. P., Elvidge, J. A., Whalley, M., *Techniques of Organic Chemistry*, Butterworth, London, 1955, p. 16.
26. Partridge, S. M., *Biochem. J.*, 1948, **32**, 238.
27. Morrison, R. I., *ibid.*, 1953, **53**, 474.

28. Isherwood, F. A., Jermyn, M. A., *ibid.*, 1949, **44**, 402.
29. *idem.*, *ibid.*, 1951, **48**, 515.
30. Brown, L., Hall, L. P., *ibid.*, 1950, **47**, 598.
31. Blainey, J. D., Yardley, H. J., *Nature*, 1956, **177**, 83.
32. Wood, T., *Biochem. J.*, 1956, **62**, 611.
33. Duncombe, W. G., Peaple, B. W. E., *Analyst*, 1957, **82**, 212.
34. Jepson, J. B., Smith, I., *Nature*, 1953, **172**, 1100.
35. Martin, A. J. P., Synge, R. L. M., *Biochem. J.*, 1941, **35**, 1358.
36. Brimley, R. C., Barrett, F. C., *Practical Chromatography*, Chapman and Hall, London, 1953, p. 26.
37. Glazko, A. J., Dill, W. A., *Analyt. Chem.*, 1953, **25**, 1782.
38. Knight, C. S., *Nature*, 1959, **184**, 1486.
39. *idem.*, *ibid.*, 1960, **188**, 739.
40. Pollard, F. H., McOmie, J. F. W., Martin, J. V., *Analyst*, 1956, **81**, 353.
41. Keller, R. A., Giddings, J. C., *Chromatographic Reviews*, Elsevier, Amsterdam, 1961, **3**, p. 1.
42. Clasper, M., Haslam, J., Mooney, E. F., *Analyst*, 1957, **82**, 101.
43. Hughes, E. B., *Modern Analytical Chemistry in Industry*, Heffer, Cambridge, 1957, p. 90.
44. Haugaard, D., Kroner, T. D., *J. Amer. Chem. Soc.*, 1948, **70**, 2135.
45. Street, H. V., Niyogi, S. K., *Analyst*, 1961, **86**, 671.
46. Pucar, Z., *J. Chromatog.*, 1960, **4**, 261; and *Chromatographic Reviews*, Elsevier, Amsterdam, 1961, **3**, p. 39.
47. Knight, C. S., *Chromatographic Reviews*, Elsevier, Amsterdam, 1962, **4**, pp. 49 and 69.
48. Weaver, V. C., *Advances in Chromatography*, Decker, New York, 1968, **7**, p. 87.

Chapter Four

Gas Chromatography

Chromatographic columns in which the moving phase is a liquid were described in Chapter 2; columns where the moving phase is a gas such as nitrogen or hydrogen will now be considered. Adsorption and partition systems are possible; the respective techniques are known as *gas-solid chromatography* (GSC) and *gas-liquid chromatography* (GLC). In gas-solid chromatography the column is packed with an adsorbent such as silica gel and the components of the mixture distribute themselves between the gas phase and the adsorbed phase, that is, on the surface of the solid. Separation is due to differences in adsorptive behaviour. In gas-liquid chromatography the column is packed with a porous solid coated with a thin layer of involatile liquid as the stationary phase. Separation is due to differences in solution behaviour. Components of the mixture distribute themselves between the gas phase and the stationary liquid phase according to their partition coefficients. The solid functions only as a support for the liquid stationary phase, enabling it to present a large surface to the gas.

Gas chromatography is arguably the most elegant and useful of the chromatographic methods, and the speed with which the technique has developed and spread is remarkable. A measure of the interest in the subject is the ever-increasing number of papers which are published; the total number of references since 1952 runs into many thousands. Knapman[1] has thoroughly documented the literature in *Gas and Liquid Chromatography Abstracts*.

Although gas-liquid chromatography is now the more important method and was predicted by Martin and

Synge[2] as early as 1941, the practical application of gas-solid chromatography antedated that of gas-liquid chromatography by several years. One of the first important accounts of gas-solid chromatography was published in 1946 by Claesson[3] who used the displacement development method (see Chapter 1) for the separation of simple mixtures of hydrocarbons on columns packed with active carbon. That was followed by the work of Phillips[4] and co-workers at Oxford who used the same method, also with active carbon columns. Phillips simplified the apparatus and showed that the method was capable of high accuracy. He separated mixtures containing as many as nine components and estimated their concentrations with an accuracy better than 2%. In 1949 Glueckauf, Barker and Kitt[5] gave an account of the enrichment of neon isotopes in which a column packed with charcoal and maintained at a low temperature was used. Cremer[6] and co-workers used the elution technique to separate mixtures of low-boiling hydrocarbons and carbon dioxide on columns of active carbon and silica gel.

In 1952 Martin and James[7] published the first account of gas-liquid chromatography and the gas-solid techniques were all but abandoned in favour of the newer method. The enormous amount of work which has now been done in the field of gas chromatography stems largely from the time of Martin and James' paper. However, for many separations, gas-solid chromatography has come back into favour.

Golay[8] in 1957 showed that ordinary packed columns in gas-liquid chromatography tended to operate well below their theoretical best performance, and was led to try coated capillaries. He found a great improvement in efficiency, and such is the rate of development that capillary columns are now in wide use for rapid or difficult analyses.

Large scale gas chromatography using large-bore columns for preparative purposes has received much attention, and for a while at least became an established technique, but the wider columns have been supplanted by smaller, packed columns operated under near analytical conditions. These are incorporated in a chromatograph that is programmed to inject numerous small samples, one after the other at appropriate intervals. A selective trapping device at the column exit collects the desired constituents of the separated mixture. In principle such an apparatus can operate indefinitely and as many samples as are necessary to form a sufficiently large aggregate may be collected. The advantage of this method over the large-scale process is that the resolving power of the column is comparable with that of a normal analytical one, whereas large-bore columns are usually markedly inferior.

Continuous gas chromatography in which a moving 'stationary' phase is used for large scale separations of simple mixtures has also received attention.

Gas chromatography is now being much used for the identification of complex substances such as polymers and biological materials[9] by pyrolysis under carefully standardized conditions, followed by separation of the gaseous products in a gas chromatograph, producing a chromatogram characteristic of the test substance.

Chromatographs are increasingly being coupled to mass spectrometers − the so-called G.C.−M.S. technique − which are powerful aids to the identification of separated compounds, particularly those present in small quantities. Again, much development has gone into the coupling of computers to gas chromatographs with two main objects. First there is the automation aspect, that is, the computer controls the chromatograph to a greater or less extent, and is responsible for monitoring the injection of the sample and mathematically analysing the chromatogram produced, presenting peak areas in digital form. Such an arrangement is particularly useful when large numbers of routine samples are involved. Sometimes the computer, if powerful enough, will use the results of an analysis for process control. Second there is the ancillary aspect where the computer is used for the analysis of the chromatogram only, particularly when the peaks are incompletely resolved (p. 265), or where there is marked baseline drift, such as occurs when there is appreciable bleed of stationary phase when operating at high temperatures (p. 190).

All of the foregoing techniques are very important, but only a selection has been included in the more detailed account that follows, since it was felt that certain topics were largely outside the scope of a book of this size.

Scope of gas chromatography

Compared with the other chromatographic method gas chromatography requires fairly complicated and expensive apparatus but the advantages more than compensate for these drawbacks. Mixtures whose separation is extremely tedious by other methods, including other chromatographic ones, can be resolved quickly and efficiently by gas chromatography. Compared with non-chromatographic techniques of similar resolving power, such as mass spectroscopy, gas chromatography is both simple and inexpensive. It is not difficult to construct home-made instruments which are very efficient. Usually the most expensive item is the recorder.

The advantages of gas chromatography depend mainly on two properties of gases, low viscosity and rapid attainment of equilibrium in certain processes. The low viscosity enables long and efficient columns to be used without serious hindrance to the moving phase. Attainment of equilibrium in the adsorption or partition process is usually rapid from the gas phase, hence high flowrates through the column may be used. Another helpful characteristic of gases is that they lend themselves readily to the determination of changes in their composition, which can be recorded continuously and automatically. It is relatively easy to vary the temperature of a column within wide limits. Finally, most gas-chromatographic columns may be used for many separations before renewal is necessary.

Clearly gas chromatography can be used only for gases and volatile substances, but the limitation imposed by volatility is not as severe as may seem at first sight. Only very small amounts of mixtures are necessary for separations, particularly when capillary columns are used, hence the vapour pressure of the substances to be separated need be only a few Newtons. Certain columns may be operated at temperatures in excess of 400°C, and thus the range of substances which fulfil the conditions is wide. Again, less volatile or less tractable compounds may have their vapour pressures increased by conversion to suitable derivatives as described on p. 226.

As is to be expected, gas chromatography lends itself particularly to the separation of organic compounds, and relatively few inorganic mixtures have been resolved. It is reasonable, however, to expect further developments in this field. Also in spite of the fact that columns may be operated at fairly high temperatures there are numerous substances — mostly inorganic — which are volatile at still higher temperatures and there is little doubt that the operating conditions of columns will be modified in order to cope with them. Nevertheless, there still remains a large number of compounds, such as proteins and dyestuffs, which cannot be volatilised, or even heated without decomposition; gas chromatography, therefore, can never replace completely the other techniques described in this book.

Technique

Many excellent commercial gas chromatographs are available of varying complexity and versatility but certain components are essential to all of them. It is these basic components that are considered in some detail in the following.

Figure 4.1 shows an arrangement suitable for gas-solid or

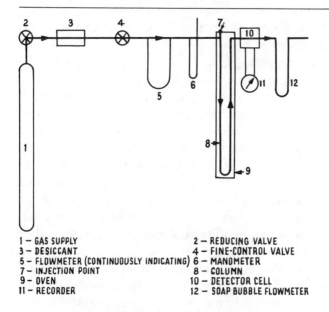

1 — GAS SUPPLY   2 — REDUCING VALVE
3 — DESICCANT   4 — FINE-CONTROL VALVE
5 — FLOWMETER (CONTINUOUSLY INDICATING) 6 — MANOMETER
7 — INJECTION POINT   8 — COLUMN
9 — OVEN   10 — DETECTOR CELL
11 — RECORDER   12 — SOAP BUBBLE FLOWMETER

Fig. 4.1. Apparatus for gas chromatography—schematic

gas-liquid chromatography provided that the elution technique is used. It may be readily adapted for use in displacement development analysis[4] by the introduction of a 'saturator' between positions 6 and 7. If a capillary column is to be used then it is necessary to incorporate a 'stream-splitting' device after the injection point 7. This is considered in more detail in the section on capillary columns (p. 233). Duplication of the flow system between positions 3 and 10 is recommended if a katharometer or gas density balance is used as the detector (see p. 195), or if there is significant bleed of stationary phase.

Gas supplies usually present little difficulty. Gases commonly employed are: nitrogen, hydrogen, helium, argon and carbon dioxide. Most of them — including mixtures such as nitrogen with hydrogen — are obtainable in large cylinders in sufficient purity to require little more than drying before use, except that uncondensable gases must be removed from carbon dioxide for the Janak absorption method. As water may be a great nuisance the gases must be thoroughly dried. The best desiccant is a molecular sieve (e.g. Linde 5A) activated at 200–300°C.

Flow control

Gas flow can be regulated accurately enough for most purposes by means of the usual high-pressure reducing valve, connected directly to the cylinder, followed by a fine control which may be a needle valve, for example, an Edward's type LB2A, or an oil-damped pressure regulator such as that manufactured by Negretti and Zambra, Type No. R/182. Additional control may be obtained when necessary by incorporating suitable buffer volumes, or chokes (in the form of capillaries) and constant flow devices such as those used by Phillips[10, 11]. Another device useful for maintaining constant flow over long periods, which has the advantage that it does not require the venting of gas to the atmosphere, is described by Hall[12].

Flow measurement

Measurement of the gas flow is most conveniently made by means of a continuously indicating flowmeter (at position 5 in Fig. 4.1). Flowrates are usually between 10 and 100 cm^3 min^{-1} and a suitable Rotameter may be used, but for more accurate work a capillary flowmeter as shown in Fig. 4.2 is convenient. Quite a good manometric fluid is the oil used in rotary vacuum pumps. It possesses a low vapour pressure and the meniscus can be clearly seen. A glass or metal capillary may be used. For the highest accuracy the device should be immersed in a thermostat bath, and calibration is against the soap-bubble flowmeter – Fig. 4.3.

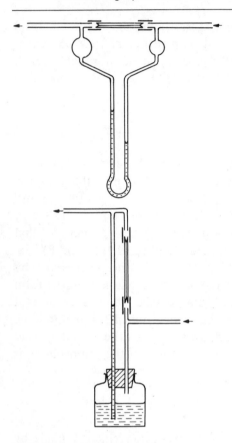

Fig. 4.2. Capillary flowmeters

This may be constructed from a 25 or 50 cm³ burette. A bubble is formed in the flowmeter by squeezing a little detergent solution into the gas stream, and it is timed between two convenient calibration marks. The method is very accurate but its drawback is that it is not continuously indicating. A correction needs to be applied for the saturated vapour pressure of the detergent solution. An alternative method of flow measurement is necessary with capillary columns and is described on p. 235.

Sample introduction

The sample may be gaseous, liquid or solid, and the method of sample introduction may differ accordingly. Sample size depends (except for preparative work) on the sensitivity of the detector; when an ionisation detector is used a liquid sample should not be greater than 0.5 mm^3, with a katharometer, not more than 10 mm^3.

Gaseous samples. A sample of gas can be injected into the carrier-gas stream through a serum cap at the top of the column by means of a hypodermic syringe; the Hamilton Teflon coated gas syringe is particularly suitable (see Fig. 4.4), or by means of a gas burette (Fig. 4.5). The latter is more accurate, but the syringe is often satisfactory if the amount of the sample is not required to be known very accurately, when, for example, proportions only are needed. The gas-burette depicted in Fig. 4.5 is a modification of one originally described by Harrison[13] and has the advantage that it can handle small sample volumes quite accurately. Sample bulb A is attached by a ball joint C to the apparatus, and air is displaced from the tubing T_4-C-T_1 by raising the mercury in B and allowing some to go to waste through T_1. Tap T_1 is now reversed and a sample of gas drawn into B. Tap T_4 is next reversed – after measurement of the volume and pressure of gas in B – and the gas sample displaced into the carrier-gas stream as rapidly as possible. Any mercury carried over can be removed through Tap T_3.

Sampling valves suitable for gaseous materials are available commercially and usually employ a by-pass loop system, an example of which is shown schematically in Fig. 4.6. With the valves in the positions shown carrier gas passes straight on to the column and sample can enter the loop; rotation of both valves through 90° enables the carrier gas to sweep the contents of the loop on to the column.

Liquid samples. Liquids are most conveniently introduced by means of a syringe, several different makes are available

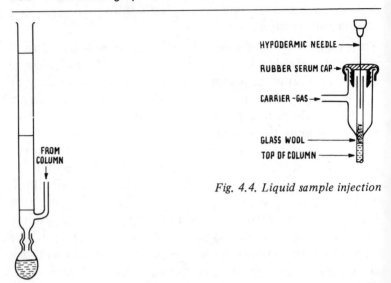

Fig. 4.4. *Liquid sample injection*

Fig. 4.3. *Soap-bubble flowmeter*

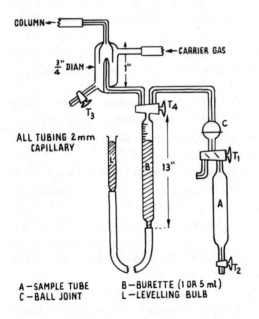

A – SAMPLE TUBE   B – BURETTE (1 OR 5 ml)
C – BALL JOINT   L – LEVELLING BULB

Fig. 4.5. *Gas sample injection*

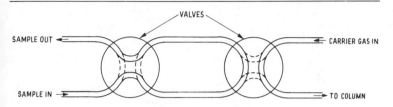

Fig. 4.6. By-pass sampling system — schematic

commercially, in a range of sizes to suit most requirements. The rubber or silicone serum cap, or disc, necessary for this method of injection may be used several times without serious leaks developing. Sometimes, however, the syringe needle will take minute cores from the disc or cap and deposit them on the top of the column where they may absorb components of the mixture and give rise to distorted peaks, or 'ghost' peaks. Very small samples can also be introduced on to the column by means of a micro-pipette, but a special column attachment is necessary to avoid interruption of the carrier-gas flow — Scott's[14] design is suitable. For very accurate work the sealed ampoule technique for solid samples may be used (see below).

Solid samples. Solid samples may comprise, among other things, greases. They can be weighed into thin glass ampoules which are placed in the gas stream and then crushed. The apparatus in Fig. 4.7 was designed by McCreadie and Williams[15]. Between 1 and 5 mg of sample are weighed into a capillary tube which is then sealed and placed in the special attachment at the top of the column. The top of the column *A* carries a C 10 ground glass socket. An ancillary tube 7.5 mm in diameter and 87 mm long has a C 10 cone at its lower end which fits into the column at *A*, and has at its upper end a C 10 socket which is connected to the carrier gas inlet. Sealed into the lower part of the ancillary tube so as to fill it almost completely is a glass rod 30 mm long. A flat surface has been ground on the rod to enable the sample capillary to fit between it and the wall of the ancillary tube,

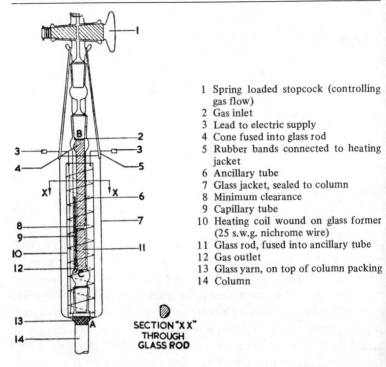

1 Spring loaded stopcock (controlling gas flow)
2 Gas inlet
3 Lead to electric supply
4 Cone fused into glass rod
5 Rubber bands connected to heating jacket
6 Ancillary tube
7 Glass jacket, sealed to column
8 Minimum clearance
9 Capillary tube
10 Heating coil wound on glass former (25 s.w.g. nichrome wire)
11 Glass rod, fused into ancillary tube
12 Gas outlet
13 Glass yarn, on top of column packing
14 Column

SECTION "X X"
THROUGH
GLASS ROD

Fig. 4.7. Sample tube crusher (after McCreadie & Williams[15])

and project about 20 mm above the rod so as to engage with the flat surface of a similar rod fused to the C 10 socket of the small adaptor (which connects the ancillary tube to the carrier-gas inlet). Carrier-gas can enter through a small hole at *B* and leave through another at *C*. There is a heater coil round the ancillary tube. The sample capillary is placed in the ancillary tube and allowed to warm up. When the correct temperature has been attained the gas inlet tube is twisted so as to turn the upper glass rod and thus break the capillary tube. An alternative method is to dissolve the solid sample in a volatile solvent and inject like a liquid sample, but the sample size cannot be measured as accurately as with the sealed tube technique.

An ingenious device that lends itself to both solid and liquid sampling is that described by Otte and Jentzsch[16] for an automatic injection system. Again, Kolb and Hoser[17] have described a useful technique for using a capsule of aluminium or gold as a microreactor for the formation of volatile derivatives, for example, of amino acids, and this capsule is used at the same time for sample transfer to the gas chromatograph.

One problem common to most injection systems is the incursion of 'memory' effects, that is, peaks may be obtained characteristic of components of samples previously injected, after an analysis has apparently been completed. Such effects may be the result of stagnant pockets of gas in the injection system, diffusion of sample upstream of the injection point during injection, or absorption by rubber or plastics fittings such as 'O' rings or serum caps. Tailing on an otherwise symmetrical peak may be another result. The remedy is to modify the geometry of the system to avoid stagnant gas and contact of the sample with materials likely to absorb it.

It is important in sample introduction that the sample itself be as concentrated as possible when it goes on to the column in the vapour state. For gaseous samples, few difficulties are experienced, provided that the charge is injected into the gas stream quickly enough, but liquid and solid samples should be made to vaporise as quickly as possible by heating the injection point by means of a small coil (*cf*. Fig. 4.7). The temperature should not be less than 50°C above the boiling point of the highest boiling component of the mixture. If, however, a very sensitive detector is used so that only very small samples are required, these precautions are not so important.

The column

The column may be glass or metal. If it is to be packed, the internal diameter should be 2 to 6 mm; and in a versatile apparatus it should be possible to vary the column length. Soft copper tubing may be used for metal columns but it can cause trouble in certain separations owing to the possibility of the formation on its inner surface of a reactive oxide that will catalyse changes in some reactive molecules. When this is not a danger, 'bright annealed seamless' copper tubing, 6 mm external diameter is suitable; it may be packed straight and then coiled round a mandrel about 20 cm diameter. It is unwise to coil more tightly owing to the possibility of disturbance of the packing and hence serious reduction of the column efficiency.

Stainless steel is a more popular material for metal columns owing to its lower reactivity as compared with copper. It may be packed and coiled in a similar way, but it is much tougher and does not work so easily. Packing is usually carried out in the already coiled column in the manner described on p. 225.

Glass columns are used in a similar way to stainless steel, although coiling is nearly always carried out before packing since high temperatures are required. The main disadvantage of glass is that it is easily broken, and its main advantage is its low reactivity. It is also possible to check the column packing visually; the evenness of the packing in metal columns must be taken on trust.

Column temperature control

The column operating temperature that is chosen will depend on the nature of the sample to be separated, but owing to the tendency of liquid stationary phases to 'bleed', that is, to be very slowly eluted from the column, it is wise to use the lowest temperature consistent with a good separation.

Operating temperatures of columns may lie anywhere between liquid nitrogen temperature (about $-196°C$) and about $500°C$, although temperatures below $0°C$ are not often employed except for the separation of the more volatile materials such as permanent gases. For very low temperatures the coiled column may be immersed in a refrigerant such as liquid nitrogen, various slushes or dry-ice-acetone mixture contained in a wide mouthed Dewar flask. Otherwise, low temperature thermostat baths may be used.

Most separations by gas chromatography are carried out at temperatures above ambient and in these circumstances an air oven with forced circulation is easily the most convenient arrangement. Astbury, Davies and Drinkwater[18] described a very efficient air thermostat in which rapid temperature changes could be made at the rate of about 20°C per minute, if required. The apparatus was, however, very elaborate and modern commercial chromatographs usually employ much simpler ovens that are, nevertheless, very efficient both for isothermal operation and when temperature programming is employed.

Ovens with linear-programmed temperature control are very useful for the separation of complex mixtures containing compounds with a wide range of boiling points. In an isothermal oven maintained at the temperature necessary for good resolution of the lower boiling components the retention times of the higher boiling materials may be unacceptably long, and the peaks very broad, whereas at the higher temperature necessary for the rapid elution of the less volatile materials the more volatile materials are incompletely resolved.

When an oven is operated isothermally it is important that the temperature does not vary more than 2°C, particularly if retention volumes are being measured, for example, for identification purposes. Variations in temperature may also affect the response of certain detectors such as the thermal conductivity cell.

For very accurate measurements of retention times such as are necessary in thermodynamic studies, thermostat baths containing water or oil, as appropriate, may be used because of the very close temperature control that is possible with them. For certain fixed temperatures columns may be enclosed in vapour jackets in which liquids of suitable boiling points are being refluxed.

Metal columns, particularly of stainless steel, may be heated by passing high current at low voltage directly through them, although it is doubtful if very constant temperatures can be maintained in this way without elaborate insulation.

Detectors

The purpose of detectors is to monitor the column effluent, measuring variations in its composition. They do not, except in the case of the so-called 'specific' detectors; usually identify the components of a mixture. So vital are detectors in gas chromatography instrumentation that almost as much time must have been devoted to their development as to that of all the other components put together.

Most detectors are of the so-called 'differential' type, that is, they give zero signal when pure carrier gas is passing through them, but when a component of a mixture is detected the signal is proportional to the concentration or mass of that component (see Fig. 4.8). 'Integral' detectors, on the other hand, give a continuous signal which is

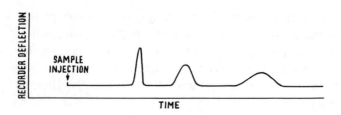

Fig. 4.8. Differential type record

proportional to the total amount of substances which have been eluted. Stepped records will be obtained as shown in Fig. 4.9. Such a record is given by the Janak absorption device or the gravimetric integral device described by Bevan and Thorburn (p. 204); each step corresponds to a different component and the step height is proportional to the amount of the particular constituent. This record may be confused with the frontal analysis or displacement development records mentioned in Chapter 1 (p. 11) which, though of similar appearance, are obtained with differential detectors.

Apart from a brief description of the Janak absorption method on page 203, only differential detectors will be considered. There are many kinds each relying on some physical property of the gas such as:

- (*i*) thermal conductivity,
- (*ii*) gas density,
- (*iii*) flame ionisation,
- (*iv*) β-ray ionisation,
  - *a*) cross-section
  - *b*) argon,
  - *c*) helium,
  - *d*) electron-capture,
  - *e*) electron-mobility,
- (*v*) photo-ionisation,
- (*vi*) microwave emission,
- (*vii*) flame emission,
- (*viii*) flame conductivity[19],
- (*ix*) rapid scanning infrared or ultraviolet absorption,
- (*x*) dielectric constant.
- (*xi*) heat of sorption.

This list is not exhaustive and only (*i*), (*ii*), (*iii*) and (*iv*) *a*)-*d*) will be described in any detail.

The requirements of a detector for gas chromatography are exacting. Among the more important are (*a*) sensitivity to

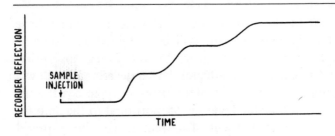

Fig. 4.9. Integral type record

very small concentrations of one gas in the presence of a high concentration of another (the carrier), (*b*) a rapid response which is also directly proportional to concentration, (*c*) a small 'dead' volume, and (*d*) a high signal-to-noise ratio.

Ionisation detectors are usually more sensitive than other types; for example, the detection limit of the thermal conductivity cell and gas density balance is about 10^{-6} moles, whereas that of the flame ionisation and β-ray (argon) detector is about 10^{-12} to 10^{-13} moles. In fact an alternative classification of detectors might be: high sensitivity — ionisation detectors, low sensitivity — others.

More recently development work has been carried out on so-called 'specific' detectors, which are instruments that respond selectively to a particular compound or class of compounds. The electron capture detector was an early example of this kind, where compounds containing strongly electron capturing atoms such as oxygen or a halogen cause a much greater response than compounds without them. The flame photometer detector is specific for compounds containing sulphur or phosphorus. When such specificity exists extremely low concentrations of the compounds sought may be detected, and the important role played by the electron capture detector in determining very low levels of chlorinated pesticide residues in wildlife is well known. Thus yet another classification is possible — specific detectors and non-specific detectors. Of the following detectors described in any detail only the electron capture and microwave plasma detectors fall in the former class.

One other classification of detectors is worth mentioning namely: concentration sensitive detectors and mass flow sensitive detectors[20]. This classification is of importance for quantitative measurements[21] (p. 215).

Thermal conductivity cell (katharometer)

In spite of frequent predictions that the thermal conductivity cell will disappear from the gas chromatography scene it still stubbornly refuses to do so, and valuable papers are still being published on its performance[22]. Certainly as more and more different detectors are developed so the katharometer will become less and less important, but it now seems doubtful if it will ever disappear completely.

The response of a katharometer depends on the thermal conductivity of the gas stream passing through it. Essentially the katharometer is a filament of a metal such as platinum (Fig. 4.10) which has a high temperature coefficient of resistance. This filament is heated by a small electric current from a six-volt accumulator. The temperature of the filament, and hence its resistance, is determined by the current and the thermal conductivity of the ambient gas. Changes in constitution of the gas change the temperature, and hence the resistance, of the hot filament. When the filament is part of a Wheatstone bridge (Fig. 4.10 (*b*)) the out-of-balance currents due to changes in the conductivity of the filament may be fed into a recorder.

The katharometer was one of the earliest detectors to be used, and for a while was one of the most popular. Provided that its limitations are realised, it is capable of satisfactory performance. In spite of its seeming simplicity, however, it is not a particularly easy instrument to make and it may prove

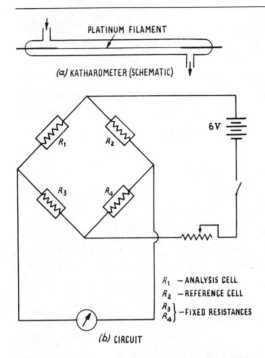

PLATINUM FILAMENT

(a) KATHAROMETER (SCHEMATIC)

6V

R_1 — ANALYSIS CELL
R_2 — REFERENCE CELL
$\left.\begin{array}{c} R_3 \\ R_4 \end{array}\right\}$ — FIXED RESISTANCES

(b) CIRCUIT

Fig. 4.10. *Thermal conductivity detector*

more convenient to buy a commercial version. A number of highly satisfactory makes are available. A simple cell is described in the Appendix.

There are several excellent accounts (*e.g.* references 21 and 23) of the characteristics of katharometers. A few disadvantages are:

(1) They are often non-linear in their response and hence require calibration.
(2) They are sensitive to changes in ambient temperature and gas flow.
(3) Their effective volume or 'dead' volume is large and they are thus unsuitable for use with capillary columns, where very small samples are used.

(4) They are not easily made sufficiently sensitive for use with very small samples and their response time is usually too long to detect substances that are eluted very rapidly.

For ordinary packed columns most of these objections are not important or can be overcome; for example, sensitivity to changes in temperature and gas flow can be minimised by incorporating the analysis and reference cells in the same metal block (see Fig. A1 and Appendix) and by making sure that the gas which flows through the two cells has passed through identical systems. This requires duplication of the apparatus between positions 3 and 10 in Fig. 4.1, one portion to be used for the separations and the other for pure carrier-gas for the reference cell.

The roles of the two portions may be reversed. It is only necessary to reverse the signal from the bridge incorporating the two thermal conductivity cells in order to give the correct response on the recorder. If the columns in the different portions are packed with different stationary phases a choice of two columns in the same apparatus is obtained.

The use of thermistors in place of wire filaments has made it possible to reduce the 'dead' volume of cells below 0.2 cm^3. Provided that they are operated at temperatures not greater than $100°C$ such katharometers may be used in conjunction with capillary columns because of their improved sensitivity and response-time compared with filament-type cells.

Katharometers depend for their operation on the thermal conductivity of the gases passing through them. It will thus be seen that the nature of the carrier-gas plays a large part in determining their performance; for example, sensitivity depends on the difference between the thermal conductivity of the carrier-gas and that of the component being sensed. Hydrogen and helium permit the highest sensitivities because the differences between their thermal conductivities and those of most of the vapours encountered in gas chromatography are substantially greater than for such carrier-gases as nitrogen and carbon dioxide.

The gas density balance

Designed by Martin[24], the gas density balance seems at first the ideal detector for gas chromatography, but difficulties of construction and lack of sensitivity compared with the ionisation detectors have prevented its more general adoption. There are now available, however, several commercial gas chromatographs which use the gas density balance or simplified versions of it. It depends for its operation on a very sensitive anemometer for detecting minute gas flows. The anemometer is located in a channel joining two other channels drilled in a copper block, one connected to the column effluent, and the other to the reference-gas supply which is pure carrier gas. While pure carrier gas emerges from the column the densities of the two streams (reference and column effluent) are the same and there is no gas flow through the anemometer; if, however, a constituent of the sample emerges from the column the density of the column effluent is changed and there will be a flow of gas through the anemometer which is proportional to the change in density. By an ingenious arrangement of the channels only pure carrier gas flows through the anemometer. It follows, from the mode of operation of this device, that the response will be a linear function of concentration and also of the relative molecular mass of the constituent. The instrument therefore requires no calibration, and can, in fact, be used to determine relative molecular masses[25].

Flame ionisation and β-ray ionisation detectors

Both the flame[26] and β-ray ionisation[27] detectors depend on the increase in current produced when eluted substances passing through them are ionised. Suitably designed, they are almost ideal for use with capillary columns and are eminently satisfactory for use with packed columns, although their high sensitivity (except the cross-section ionisation detector) may prove an embarrassment. Their 'dead' volume may be extremely small, their response is almost instantaneous, they are very sensitive and they are stable to fluctuations in the flowrate of the gas, and to changes of temperature. Finally, they give a response which is a linear function of concentration – at least for the concentrations normally met in gas chromatography. A comprehensive review of ionisation methods for the analysis of gases has been given by Lovelock[28].

Flame ionisation detector

In the flame ionisation detector shown schematically in Fig. 4.11 hydrogen is used as the carrier gas, or is added to it, and is burned in a small metal jet, such as a hypodermic needle cut square. The jet forms one electrode – usually the negative – of the cell and the other, which may be a piece of brass or platinum wire, is mounted at some point near the tip of the flame. The potential difference across the electrodes is about 200 V. Pure hydrogen or hydrogen/carrier gas mixture gives rise to a small background signal which may be offset

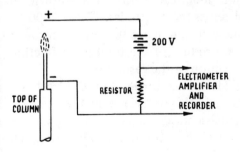

Fig. 4.11. Flame ionisation detector (schematic)

electrically, and the signal which is recorded arises from ionisation of substances in the flame, including flammable as well as non-flammable compounds. The mechanism of ion production is not completely clear, but for organic substances, ions may originate from carbon aggregates which ionise relatively easily (approximately 415 kJ mol^{-1} [4.3 eV]). Response of the detector to organic substances also depends on the 'carbon number' of the compound. A design for a flame ionisation detector is included in the Appendix.

According to Condon[29] the following substances are not detected because of their high ionisation potentials:

the noble gases, H_2, O_2, N_2, $SiCl_4$, $SiHCl_3$, SiF_4, H_2S, SO_2, COS, CS_2, NH_3, NO, NO_2, N_2O, CO, CO_2, H_2O.

About the only organic compound not detected is formic acid. An advantage of the flame ionisation detector is that it does not indicate the presence of water thus enabling the analysis of aqueous solutions to be performed.

Cross-section ionisation detector

The cross-section detector was the first ionisation method to be used and is possibly the least sensitive. A source of ionising radiation such as ^{90}Sr is contained in a brass vessel similar to (though not identical with) that shown in Fig. 4.12. The potential across the electrodes is 300–1000 V. Production of ions in the cell, and hence the current through it, is proportional to the product of the total mole fraction and the ionisation cross-section of the gas mixture.

Any gas may be used as the carrier but the best are hydrogen or helium since both give relatively few ions on account of their small cross-sections. Large molecules such as are likely to be separated by gas chromatography are much more readily ionised and therefore increase the current through the detector.

Advantages of the detector are its ability to respond to all gases and vapours, in any concentration and in any carrier-gas, but the lower limit of detection is about 10^{-5} moles.

Argon ionisation detector

The argon ionisation detector (Fig. 4.12) is in many respects similar to the cross-section ionisation detector, but is far more sensitive. The cell is a stainless steel or brass vessel of small volume, which acts as the cathode; the anode is of similar material. The applied potential is usually between 600 and 1200 V; the higher the potential the greater the sensitivity. A β-emitter such as ^{90}Sr, ^{85}Kr or tritium is sealed into the vessel.

Ionisation of substances in the stream of argon carrier-gas occurs mainly as a result of collisions with metastable argon atoms. Production of metastable atoms appears to occur mainly as follows: electrons from the radioactive source ionise argon atoms (and, to some extent any other atoms or molecules which happen to be present) and the electrons thus released are further accelerated by the applied field so that other argon atoms with which they collide are excited to their metastable state. A single electron can produce several excited atoms. Metastable argon atoms have an excitation level of 1129 kJ mol^{-1}; hence, collision of substances with

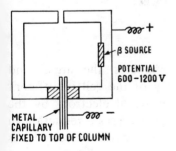

Fig. 4.12. β-ray ionisation detector (schematic)

lower ionisation potentials results in the further production of ions and this increases the conductivity of the gas. Clearly, substances with ionisation potentials greater than 1129 kJ mol^{-1} will not be detected, or only give rise to a poor signal; among them are H_2, N_2, O_2, CO_2, $CO(CN)_2$, H_2O and fluorocarbons, which are not detected, and CH_4, C_2H_6, CH_3CN and C_2H_5CN, which give a small response. The difficulty of detecting these substances can be overcome to some extent by an ingenious method described by Willis[30]. Advantage is taken of the fact that such substances reduce the ion current produced in the detector by an organic vapour such as ethylene, and they can therefore be detected in the presence of such a vapour by their 'negative' peaks.

Helium ionisation detector

Berry[31] has described an ionisation detector which depends on the use of highly purified helium as the carrier gas. The construction and mode of operation are similar to those of the argon detector and the sensitivity is comparable. However, helium has an excitation level of 1910 kJ mol^{-1} which means that gases not sensed in the argon detector are ionised and therefore give good signals. Even neon (ionisation potential 2074 kJ mol^{-1}) is ionised to some extent in this detector, possibly by a two-stage process.

Electron-capture detector

The electron-capture detector resembles the argon and helium detectors, but the geometry of the cell is different and the signal depends on the capture of electrons by the various substances being sensed which causes a reduction in the ion current. The applied voltage across the electrodes is only about 20 V. Particular advantages of this detector are its very high sensitivity and specificity for molecules such as oxygen, halogens, oxygen-containing and halogen-containing compounds, which have high electron affinities. The response of weakly electron-capturing compounds, such as hydrocarbons and ethers, can be eliminated by increasing the applied potential. This means that the detector can be made selective in its response.

Microwave plasma detector

The microwave plasma detector, or MPD, is a fairly recent development and may be regarded as a specific detector, but with the considerable advantage that it may be tuned to any element – or number of elements – including those most frequently sought by chromatographers, namely, C, H, N, P, S and halogens. It is basically an atomic emission device, the emission resulting from the passage of eluted components through a microwave-sustained helium plasma discharge that causes them to be completely dissociated into their constituent atoms.

In a particular commercial instrument the column effluent is divided into two equal streams, one going to a flame ionisation detector, the other to the plasma tube. The 'interfacing' with the chromatograph is thus relatively simple.

Although expensive, the MPD would seem to offer substantial advantages over most other detectors; in addition its sensitivity is comparable to that of the flame ionisation detector.

Integrating detectors

The first detector described in association with gas-liquid chromatography was an integral detector devised by Martin and James[7] in which volatile fatty acids were titrated against sodium hydroxide solution as they emerged from the column by an automatic burette. This device was clearly of limited application and hence was not generally adopted, but in common with the other instruments mentioned below, it did directly measure the amount of substance — other than the carrier gas — passing through it and therefore was particularly suitable for quantitative work.

In the gravimetric integral detector devised by Bevan and Thorburn[32] the column effluent is led into an adsorption cell where it is taken up by the material lining the cell and hence increases its weight. The cell is supported by an auto-recording gravimetric balance with a sensitivity of about 10^{-7} g; thus separated substances are detected and measured absolutely according to their mass.

The Janak[33] absorption method for the detection and estimation of the separated components of a mixture is simple and relies on the measurement of gas volumes. It is suitable for gases with relatively low boiling points such as the low molecular weight hydrocarbons. The column is operated in the usual way but the carrier gas is carbon dioxide, which, after passage through the column, is absorbed by a strong solution of potassium hydroxide. Any non-absorbed gases pass into a gas burette where their volume is measured at constant pressure. The non-absorbed gases are, of course, the components of the original mixture. If the volume of gas in the burette is plotted against time a stepped curve is obtained which is similar to that shown in Fig. 4.9. This is an 'integral' record as already mentioned, and might be confused with the record obtained during a frontal or displacement development analysis. In Fig. 4.9 the heights of the steps are proportional to the amounts of the particular constituent in the original mixture. The identity of a particular component can often be inferred from the time taken for it to emerge from the column, measured from the injection time of the original mixture.

Recorders

Choice of recorder depends very largely on the funds available. Practically any potentiometric recorder (range about 1 mV) will do for most of the methods of detection including thermal conductivity, flame ionisation, beta-ray ionisation and gas density, only minor modifications being necessary. An additional voltage amplifier is necessary with the gas-density balance, and the ionisation detectors described require electrometer amplifiers.

Identification of the components of a mixture

The most reliable way of identifying the components of a mixture separated by chromatography is to trap them as they emerge from the column, and to identify each separately by techniques such as infrared and mass spectroscopy, but see below (p. 212). Needless to say, such a procedure is tedious and, when routine separations are being performed on mixtures whose composition is known to vary only between fairly narrow limits, unnecessary. Other methods of identifying components are based on chromatographic behaviour. In routine analysis, provided that the experimental conditions are always the same, the identity of the peaks can be inferred from previous separations. The method depends on the fact that for a given set of experimental conditions, components will always behave in the same way. Another way is to make up synthetic mixtures and compare their behaviour with that of the unknown. An unfamiliar peak can sometimes be identified by adding to the mixture some of a pure component thought to be identical with it. If there is an increase in size of the unknown peak, then there is a good chance that the pure substance added is identical. Before describing the chromatographic methods of identification in more detail it will first be necessary to define the various retention volumes which are used. The treatment applies particularly to gas-liquid chromatography, but provided that more or less symmetrical peaks are obtained, it can also be applied to gas-solid chromatography. Caution is, however, necessary in the latter case because of the inconstancy of adsorbent behaviour that may occur from one batch of material to another. The presentation of retention data is discussed in references 34 and 35.

Retention volumes

Retention volumes are expressed in cm^3 and reference is made to Fig. 4.13. When a substance is injected on to a chromatographic column, a small volume of air is usually introduced. It is assumed that air is not retarded by the stationary phase at normal operating temperatures, therefore

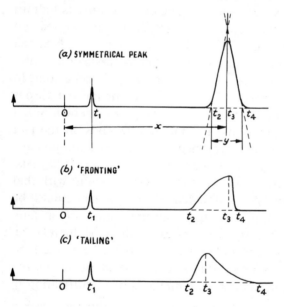

Fig. 4.13. *Elution peaks*

the volume of carrier-gas required to elute the air is a measure of the volume of the apparatus, between the injection point and the detector, not occupied by column packing. This volume, known as the '*gas hold-up*' V_m, is given by:

$$V_m = F \cdot Ot_1$$

where F is the flowrate (cm^3 min^{-1}) of the carrier-gas at the outlet pressure and temperature of the column, and Ot_1 is the time taken for the air-peak maximum to appear.

An air peak will not appear with some instruments, such as the flame ionisation detector. Methane, which is detected, is usually retained to a negligible degree by the stationary phase (except in gas-solid chromatography at low temperatures) and may be injected separately to determine the gas hold-up with reasonable accuracy. However, Thombs[36] has reported that pulses of noble gases produce responses from the flame ionisation detector and hence may be used for gas hold-up measurements.

The *uncorrected retention volume* V_R of the substance is given by:

$$V_R = F \cdot Ot_3$$

where Ot_3 is the time taken for the emergence of the peak maximum of the substance. This includes the gas hold-up of the apparatus and making allowance for this we obtain the *adjusted retention volume* V_R' which is given by:

$$V_R' = F \cdot t_1 t_3$$

where $t_1 t_3$ is the time taken for the emergence of the peak maximum measured from the air-peak maximum.

There is a pressure gradient down the column and it is therefore necessary to introduce the so-called 'compressibility factor' (j) of Martin and James[7] which is given by:

$$j = \frac{3(p_i^2/p_0^2 - 1)}{2(p_i^3/p_0^3 - 1)}$$

where p_i is the pressure of carrier gas at the column inlet, and p_o the pressure at the column outlet. Hence is obtained the *net retention volume* V_N given by:

$$V_N = j V_R{}'$$

Finally, the

specific retention volume V_g, is the net retention volume of the substance per g of stationary phase, at $0°C$, thus:

$$V_g = V_N \cdot \frac{273}{Tw}$$

where T is the temperature of the column in degrees Kelvin and w is the weight in g of stationary phase in the column.

Relative retention volumes are obtained by comparing the retention volumes ($V_R{}'$, V_N, or V_g) of the solute under consideration with some standard solute whose behaviour on the particular column in use is precisely known. Relative retention volumes are conveniently expressed as ratios: for example

$$\frac{V_N \, (\text{Standard})}{V_N \, (\text{Test})}$$

Specific, net and relative retention volumes may be used for identification purposes.

Use of specific retention volumes

The specific retention volume of a solute is characteristic and may be used to identify peaks in a chromatogram. It is dependent only on the stationary phase in use in the column, and the temperature provided that the stationary phase is a single, pure compound. If it is a mixture, for example, a silicone, there may be poor reproducibility from batch to batch. In view of the temperature dependence it is obviously important that the value of V_g used is the correct one for the particular column conditions. It is possible to calculate the value of V_g for different temperatures by means of an Antoine equation:

$$\log V_g = A + \frac{B}{t + C}$$

where t is the column temperature in degrees centigrade and A, B and C are constants which can be evaluated graphically from gas-liquid chromatography data. Ambrose and Purnell[37] have determined the constants for a number of organic compounds, and the temperature range over which the values are valid is quoted. Few V_g values have been determined, probably because of the care needed to obtain accurate results. Adlard, Khan and Whitham[38] have made very careful measurements for benzene and cyclohexane on columns of Celite coated with 20% dinonyl phthalate, in the temperature range 48–110°C. In practice there is a tendency to use relative retention volumes.

Relative retention volumes

Relative retention volumes may be used for identification purposes provided that the experimental conditions are as nearly identical as possible for the test substances and the standard. This may be ensured by adding the standard substance to the original mixture. Alternatively, separate successive runs may be made with the standard and then the mixture. The method is to compare the retention volumes (or times) of the other peaks with those of the standard and see if the ratio corresponds to that of a known compound with the standard. Like V_g values, relative retention volumes are valid for one temperature only. Variations of the relative retention volume method have been described by Kováts[39] and Smith[40] although Kováts' recommendations have been the more generally adopted.

In Kováts' method, which is based on the nearly linear relationship between log V_R' and the carbon number in homologous series, normal alkanes are used as reference substances and the relative retention of the test substance, B, is expressed in terms of its so-called Rentention Index, or *RI*. This is based on the Retention Indices of the two consecutive normal alkanes, nA_a and nA_{a+1}, that are eluted either side of it in the chromatogram, as shown in Fig. 4.14:—

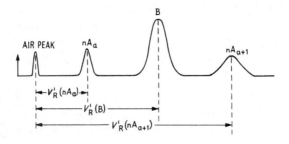

Fig. 4.14. Derivation of retention indices

the arbitrary *RI* that is assigned to each normal alkane is 100 times its carbon number, thus if nA_a and nA_{a+1} correspond to n-butane and n-pentane, their *RIs* are 400 and 500 respectively, and the *RI* of B would be some value between 400 and 500.

A logarithmic scale is used, as mentioned above, and the *RI* of substance B, *RI*(B) is given as follows:

$$RI(B) = 100A + 100 \frac{\log V'_R(B) - \log V'_R(nA_a)}{\log V'_R(nA_{a+1}) - \log V'_R(nA_a)}$$

A further condition is that

$$V_R'(nA_{a+1}) > V_R'(B) > V_R'(nA_a).$$

In the above A is the carbon number of the alkane nA_a, and $V_R'(B)$, $V_R'(nA_a)$ and $V_R'(nA_{a+1})$ are the adjusted retention volumes of substance B, n-alkane nA_a and n-alkane nA_{a+1} respectively.

On application of the Kováts' system a number of useful points are seen to emerge, including the following: the RIs of a given substance are nearly constant in all non-polar stationary phases.

RIs of members of an homologous series increase by 100 units for each additional CH_2 group.

The RI of a given substance in a non-polar stationary phase changes on passing to a polar one, according to the nature of the functional groups in the given substance. Thus it is possible to calculate the RI of a substance in a polar stationary liquid provided that its RI in a non-polar stationary liquid is known and also the change (increase or decrease) due to the functional group, or groups, are known, and vice versa. Similarly it follows that the change in RI from a non-polar to a polar stationary phase can give an indication of the type of functional group present.

By applying a FORTRAN computer programme[41] RIs may be calculated without the need for bracketing each compound in a mixture with two normal alkanes. The programme can also take into consideration the variation of the retention time with amount of substance injected (see below).

Log V_N, and log V_g plots

By plotting log V_N or log V_g against the 'carbon number' for a homologous series nearly straight lines are often obtained. This is true for many different types of compound, including paraffins, alcohols, fatty acids, esters and others. If an unidentified peak is suspected of belonging to a particular homologous series it becomes a simple matter to identify it, provided that the log plot is available. Again, such plots are valid for one temperature only. It is possible, however, to convert the V_N or V_g values to other temperatures by means of Antoine equations. Phillips, Littlewood and Price[10] have plotted experimental values of log V_N against $10^4/T$ for a number of organic compounds, and have obtained straight lines.

Sources of error in the determination of retention volumes

One of the most troublesome sources of error in the determination of retention volumes is that due to sample size. It is often found that the retention volume of a particular substance varies with the size of the sample injected, possibly due to non-linearity of the partition isotherm, or the incursion of adsorption effects. In either case the elution peaks will tend to be asymmetric. Systems are known in gas-liquid chromatography where fronting is observed below a certain column temperature and tailing above it, usually when the partition coefficient is large[42]. With asymmetric peaks more consistent results may often be obtained by measuring the retention volume to the beginning of the emergence of the peak (t_2) (Fig. 4.13 (b)), when 'fronting' occurs, and the final part of the peak in the case of 'tailing' (t_4 in Fig. 4.13 (c)). 'Fronting' and 'tailing' effects may often be avoided if a small enough sample is used.

That retention volumes must always be used with caution is underlined by the interesting example given later in which the order of elution is changed by varying the proportion of the stationary phase (p. 224).

Gas chromatography — mass spectrometry

When the trapping of samples is difficult as in the case of chromatography on capillary columns, or tedious when very large numbers of components (greater than 100, say) are involved, it is becoming increasingly common to link mass spectrometers directly to gas chromatographs so that eluted components may have their mass spectra recorded at about the same time as the usual peaks are recorded in the chromatogram. Mass spectrometers are particularly valuable for identifying single substances, but are much less satisfactory for mixtures. The coupling of a chromatograph, which is primarily a separation device, with a mass spectrometer is therefore an extremely powerful combination, although a rather expensive one.

The main problem in associating a gas chromatograph (GC) with a mass spectrometer (MS) is the so-called interface between the instruments. The proportion of the total amount of a component eluted that may be directly introduced into the MS is determined by the pressure limits acceptable in the ion-source. The total effluent from a capillary column, provided that its bore is small enough, can be directed into the MS, but a small fraction only is acceptable from packed columns, hence a stream-splitting device, similar to that described for the injection system for capillary columns (p. 234) will be necessary. A further complication, however, is the amount of carrier gas – invariably helium for GC–MS combinations – that may be present and that will 'dilute' the mass spectrum of the components under examination. The maximum flowrate tolerable for helium is between 0.1 and 2.0 cm^3 min^{-1}. The problem may be eased by passing the column effluent, or whatever fraction of it that may be acceptable, through a molecular separator[43]. A Biemann-type separator is shown in Fig. 4.15.

The effluent from the GC passes through a porous glass tube, whose walls are permeable only to the carrier gas, which is continuously pumped externally. The effect is to increase the concentration of the sample relative to that of the carrier gas. It may be important to heat the complete line from the GC

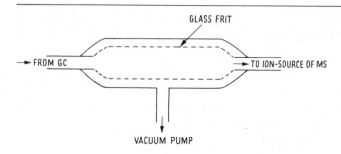

Fig. 4.15. Biemann separator — schematic

to the MS, including the separator, to avoid adsorption effects and hence distortion of peaks, particularly for materials of higher molecular mass. In this connection platinum capillaries have been recommended for direct coupling of glass capillary columns to the MS[44]. Advantages are favourable surface behaviour, good mechanical strength and ease of sealing to the glass capillary and to the steel connection to the MS source.

It is possible to operate the MS as a specific detector by selecting, and locking on to, the mass number of an ion known to be characteristic of a particular substance and determining only that mass. Similarly, by selecting a fragmentation ion common to all the components of a mixture the MS may be made to function as a non-specific detector and to display the variation of the mass of the common ion with time on an oscilloscope, or on a potentiometric recorder. The trace then resembles a conventional chromatogram.

Anthony and Brooks[45] have further refined the GC—MS technique for the identification of steroids, terpenes and other molecules by deuteration of the compounds on the column itself, followed by measurement of the mass spectra. These provide valuable supplementary information on structure and fragmentation and assist in characterisation of the molecules being examined.

Quantitative analysis

It is interesting to recall that the earliest papers on gas-solid chromatography[3, 4] and gas-liquid chromatography[7] were concerned specifically with quantitative aspects of the technique; accuracies and precisions in analysis were achieved that compare favourably with modern practice. Of the total literature on gas chromatography, however, comparatively little has been published on quantitative analysis, but Guiochon and co-workers[46] have given a valuable review of precision and accuracy in gas chromatography and conclude that the technique is, at least potentially, one of the most accurate analytical methods now available.

When an integral detector such as that described by James and Martin[7], Janak[33] or Bevan and Thorburn (gravimetric integral detector) is used it is easy to estimate the quantity of a given compound in a mixture since this is determined by direct measurement of a titration volume, a volume of gas or a mass of material. Unfortunately most modern detectors are of the differential kind and quantitative measurement is based on the relationship between the quantity of the given compound introduced into the chromatograph and either the area or height of the corresponding peak in the chromatogram; calibration is therefore essential even if the detector response is predictable as in the case of the gas density balance. In the most favourable cases there is a linear relationship between the peak height or area and the amount injected, but with most detectors linearity becomes the less likely the greater the concentration of the component.

Errors may arise in several ways, but principally from the instrumentation, the method of introduction of the sample and from the method of measuring the peak height or area.

Sample introduction by injection with a syringe is cheap and convenient, but less precise than by sampling valve or sealed capsule techniques. The syringe method is probably most precise with gases on account of the relatively large volumes involved, but it is doubtful if the precision in this most favourable case is better than about 2%, while with liquids it may be as great as 5%. Whatever method is used it is important to ensure rapid and complete vaporisation of the sample, absence of stagnant pockets of gas in the injection system and complete absence of leaks in the transfer lines.

The detector response may, or may not, be a linear function of the mass or concentration (see p. 194) of material passing through it; it is still essential in either case to carry out careful calibrations. Katharometers, for example, are sensitive to flowrate and temperature, whereas the beta-ray ionisation detectors are little affected by either.

Peak height measurement probably gives more accurate results than measurement of peak area, particularly when the peaks are relatively tall and narrow, but very accurate control of temperature and flowrate is essential. Heights can be measured from the chromatogram (Fig. 4.16) or directly from the detector signal. Peak area measurement does seem to be more popular when the peak is shallower and wider, as in Fig. 4.16 (*b*). With symmetrical peaks a fairly accurate estimate of their areas is obtained by multiplying the peak height by the peak width at half peak height (Fig. 4.16 (*b*)). The area thus obtained is 0.94 of the true area. With unsymmetrical peaks the area may be measured by means of

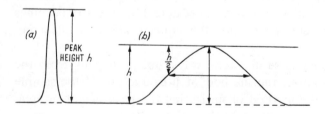

Fig. 4.16. Measurement of peak height and peak area

a planimeter, or an integrator with the recording system, although an additional instrumental error may be introduced. A quite accurate method is to cut out the peak from the recorder chart and weigh it. The accuracy is sometimes improved by photocopying the chromatogram and cutting out from the copy since this reduces the effects of variations in the recorder strip arising from the printed lines, the weight of the paper or even the recorder trace itself.

Calibration is effected by introducing accurately known amounts of the substances of interest into the chromatograph and comparing them with the peak height or area in the chromatogram that results. Response factors may then be calculated, that is, unit height or area is expressed in terms of the amount of substance present. Non-linearity of detector response can be allowed for by careful calibration.

Deans[47] compares three different methods of obtaining quantitative results, normalisation, constant volume injection and internal standard. In the normalisation method all the peaks in the chromatogram are measured and multiplied by their response factors; the results are then normalised to give percentage concentrations. The main advantage is that no sample preparation is required and small changes in the operating conditions are usually unimportant, but disadvantages include the fact that errors arise if any components of the mixture do not give detectable or measurable peaks. Also, all peaks have to be identified and response factors known — whether they are of analytical interest or not.

The constant volume injection method is simpler; repeatable volumes of the sample are injected, peaks of interest are measured and converted by means of their response factors to provide the required analytical data. Again, no sample preparation is required, and only peaks of interest need to be separated. Among disadvantages may be mentioned the difficulty of achieving good repeatability of sample volumes.

Finally, in the internal standard method, a known amount of a marker compound is added to the sample before injection. Ratios of peaks of interest are then measured and converted, by means of their response factors, to concentrations relative to the marker. This method combines some of the advantages of the other two methods described; only peaks of interest need to be separated, and small changes in the operating conditions are unimportant because response factors relative to the marker are measured. The main disadvantage is that an accurately known concentration of the marker must be added to the sample.

Sources of error are fully discussed by Deans[47].

Gas-liquid chromatography

Preparation of a column for gas-liquid chromatography involves the choice of the solid support, the stationary liquid phase, and the type of column (capillary columns are dealt with separately later). Points to consider in connection with column type (metal or glass) have already been mentioned (p. 189).

Solid support

The most popular solid supports are diatomaceous earths (kieselguhr) such as Celite and Embacel, and the less expensive crushed firebrick such as Stermachol and Johns-Manville C.22. These materials are supplied in a variety of particle sizes and can be obtained acid or alkali washed, or both. Both kieselguhr and firebrick are siliceous, highly porous solids; their chief difference, apart from chemical composition, appears to be in their surface areas. Measurements[48] made on Celite and a crushed firebrick by means of krypton adsorption at liquid nitrogen temperature, gave a value of about 0.5 m^2 g^{-1} for the former and 7.5 m^2 g^{-1} for the latter. This difference is further emphasised in a packed column because Celite is much less dense than firebrick. Ettre[49] has given the following figures for surface areas: firebrick 4.14, Chromosorb 2.86, Chromosorb W 1.41, and Celite 1.14 m^2 g^{-1}.

When the large amount of liquid, which these solids can absorb without becoming noticeably sticky, is considered, the figures suggest the presence of relatively large pores compared with those in adsorbents like silica gel and alumina. Large pores can, in fact, be seen in kieselguhr under an optical microscope. Electron microscopy further reveals that the diatoms, which make up the macroporous structure, are themselves porous. These 'micropores' (still large on a molecular scale) in the plate-like diatom skeletons are extremely regular in size. It may be that the stationary phases actually form minute lenses of liquid in the micropores. This raises the possibility that the stationary phase is not spread in an even film over the surface of the support, but that areas of bare, or only thinly coated, support are accessible to gases. However, an investigation by Drew and Bens[50] using a scanning electron microscope showed little apparent difference between the coated and the uncoated support, and there appeared to be little evidence for pore filling or lens formation by the liquid stationary phase.

Catalytic effects are usually less with Celite than with firebrick, though acid washing may diminish such effects with both solids, presumably by removing iron. Tailing is more likely to occur with firebrick[51, 52], possibly on account ot its more polar character and its larger surface area, but there is often tailing, though less marked, on Celite. Treatment of the solids with the vapour of dimethyldichlorosilane (DMCS) or hexamethyldisilazane (HMDS) prior to coating with the stationary phase helps to eliminate these undesirable effects. A comparison of the adsorption characteristics of some solid supports has been made by Bens[51].

Other materials that have been used as solid supports

include; granular polytetrafluoroethylene; Kel F (a chloro-fluoro polymer); porous polymers of the polystyrene type (see also p. 255); fused glass, fused silica and fused alumina spheres; sodium chloride crystals; metal helices. The polymeric materials are useful in providing non-polar surfaces that exhibit reduced adsorption effects for polar substances. They are sometimes difficult to pack, however, owing to their tendency to acquire an electrostatic charge, but prior cooling to $-78°C$ or less helps in the production of more efficiently packed columns than can be achieved at room temperature.

The fused or crystalline materials are mostly non-porous — although they may have rough surfaces which help to maintain an even coating of liquid stationary phase — and usually cannot hold more than about 3% of their weight of liquid phase, but this is not necessarily a disadvantage (Littlewood[53]).

Contrary to accepted practice, the use of active alumina has been advocated by Halasz and Wegner[54] as a support for polar liquids like triethylene glycol and β-β'-oxydipropionitrile. It is claimed that a more efficient column is obtained than when an inactive support is used. A significant feature is that retention volumes (p. 206) decrease with increasing proportion of liquid phase suggesting that some adsorption occurs on the support when it is coated with the smaller amounts of liquid.

The above example draws attention to the fact that, just as it is sometimes difficult to distinguish between adsorption and partition in liquid chromatography so there are gas chromatographic packings that exhibit the properties of both solids and liquids for chromatographic purposes. Uncoated porous polystyrene beads and solids like Celite with the stationary phase chemically bound to the surface are other examples. There is a further discussion of such materials under gas-solid chromatography (p. 238). It is also worth mentioning that suppliers of solid supports and stationary phases usually provide very informative catalogues describing their products.

Particle size

Particle sizes used for solid supports usually fall within the range 30 to 120 mesh B.S.S. Narrow fractions such as 80 to 100 or 100 to 120 mesh are to be preferred because they lead to more even packing in the column and there is less tendency for clogging to occur. Proprietary brands of kieselguhr are supplied ready for use but if crushed firebrick is preferred it is important to remove 'fines' after sieving, since the very small particles can interfere with separations either by clogging the column or distorting the chromatographic bands. 'Fines' can be removed by sedimentation in a large measuring cylinder filled with water. After a preliminary wash with water the solid is dried in an oven. When dry it is poured into the water in the cylinder and allowed just to settle, while the supernate containing the very fine particles is decanted. Finally, the solid is dried thoroughly at $150°-200°C$ before use.

Liquid stationary phase

The large number of liquids suitable for use as stationary phases in gas-liquid chromatography provides a somewhat bewildering choice, but it will be found that surprisingly few are necessary for the separation of a very wide range of chemical compounds. Table 4.1 includes some of the more popular substances in general use.

Maximum operating temperatures may have to be lower than those given in the table, particularly if very sensitive detectors are used, since the presence of stationary phase in the effluent gas stream may give an appreciable background

Table 4.1   *Stationary phases for gas-liquid chromatography*

Substance	Maximum operating temperature, °C	Used for separating
squalane	120	hydrocarbons, general application
Apiezon M	150 ⎫	hydrocarbons, general application
Apiezon L	230 ⎭	
benzyl biphenyl	120 (m.p. 40–60)	hydrocarbons, aromatic compounds, halogen compounds
dimethylformamide	20	low boiling paraffins and olefins
n-hexadecane	50	hydrocarbons
tritolyl phosphate	120	chlorinated hydrocarbons
dinonyl phthalate	120	general application
diglycerol	150	alcohols, selectively retains polar materials
polyethylene glycol	100	alcohols, ketones
polyethylene glycol adipate	150	esters, fatty acids
silicone oils and gums	depending on the constitution, up to 400	high temperature work, general application, sterols

signal. This may be partially eliminated by a dual column arrangement – cf. p. 197. More comprehensive lists are given in references [55] and [56]. Choice of the particular liquid depends on the column temperature to be used and on the chemical nature of the compounds to be separated. If tailing occurs it usually means that adsorption is taking place on the solid support and it can often be controlled by the use of a more polar stationary phase or by the pretreatment of the support with such substances as DMCS or HMDS (see p. 219). Knight[52] says that tailing can also be controlled by the addition of a strongly adsorbed material such as water to the carrier gas, but, as already stated, water is not usually a desirable substance to have present in gas chromatography unless a flame-ionisation detector is used.

Porter and Johnson[57] have used heptane, octane and acetone as stationary phases in columns maintained at −78° for the separation of many low-boiling gases including hydrogen, nitrogen, oxygen, argon, carbon monoxide and light hydrocarbons. This seems to be a very useful alternative to the use of gas-solid chromatography for such separations. For very high temperatures, up to 500°C, a polycarborane-siloxane stationary phase has been developed commercially (Olin Corporation) that possesses negligible vapour pressure and is thermally stable at this high temperature. The use of chemically bound stationary phases already mentioned (see also p. 99) permits the operation of columns without significant bleeding at temperatures up to the dissociation temperature, but these temperatures are disappointingly low — 150–200°C in most cases.

There are some very selective stationary phases: silver nitrate dissolved in polyethylene glycol forms loose adducts with olefins and is specific for such compounds provided that the temperature is not too high. Interesting compounds used by Phillips and his co-workers[58] are the *N*-dodecylsalicyl-aldimines of nickel, palladium, platinum and copper, and also the n-octyl glyoximes of nickel, palladium and platinum which selectively retain molecules which can act as ligands to the transition metals, such as amines, ketones, alcohols and molecules containing olefinic and acetylenic bonds. Again, tri-*o*-thymotide dissolved in tritolyl phosphate will selectively retard straight chain organic compounds relative to those with branched chains. Dicarboxylrhodium(II)trifluoroacetyl (+)camphorate has been used, dissolved in squalane, for purity determinations and separation of olefins[59], while dimeric rhodium(II)benzoate in squalane reacts reversibly with substances such as ethers, ketones and esters capable of lone pair donation, and considerably increases their retention volumes. Other transition metal compounds have also been investigated such as vanadium(II) and manganese(II) chlorides.

Mixtures of liquids may be used but in the cases so far investigated the effect is the same as that given by two columns made up from the separate constituents of the mixture. Static measurements made by Freeguard and Stock[60] show that the isotherms obtained with a mixture of dinonyl phthalate and squalane are almost exactly those which would be expected on the assumption that the two components sorb the vapours presented to them, independently of each other. Molten salt mixtures have been used as stationary phases; the eutectic mixture of sodium nitrate, potassium nitrate and lithium nitrate[61] has been used for the separation of organic mixtures and appears to behave like any other stationary phase.

Proportion of liquid phase

Although the proportion of stationary phase does have an effect on the performance of a column it is usually not critical (but see below). Proportions of stationary liquid vary from about 1 to 30% by weight of the coated support; 20% is common. Use of the smaller proportions of stationary phase, 1 to 5%, has the advantage that separation times are shorter, and that sorption equilibria are more rapidly established than for thicker films. There is a distinct danger, however, of adsorption effects on the solid support, particularly if firebrick is used. Again there is the possibility of adsorption at the liquid-gas interface becoming important when small amounts of stationary liquid are used. Martin[62] suggests that this effect will be most pronounced when the liquid is highly polar and the surface area of the support is high. The effect will become less important with increasing proportion of liquid phase because of the smaller surface-to-volume ratio. This may be the explanation of the fact that chlorinated phenols emerge from a column, with 10% Apiezon grease as the stationary phase, in the order of their boiling points, but with 30% grease they emerge in the order of their polarities[63]. The 'mass transfer' term in the van

Deemter equation is also smaller for smaller film thicknesses (see p. 269).

Preparation of the coated solid

To prepare the coated solid a known amount of the stationary phase is dissolved in a volatile solvent such as ether and a known weight of the support material stirred in. The solvent is removed by gentle warming, with thorough agitation of the mixture to ensure even deposition of the stationary phase. Finally, the coated solid is dried in an air oven for several hours. It is necessary to purge the column with the carrier gas overnight at a temperature higher than the operating temperature to remove the last traces of solvent and volatile impurities. If the amount of stationary phase is required to be known accurately a weighed amount of the coated solid can be extracted with a solvent and then reweighed. Some stationary phases are hygroscopic and should be stored in airtight containers.

Filling the column

A correctly packed column is vital for efficient operation. Great care is necessary to ensure an even filling. The following method may be adopted: while the column is gently bumped on the floor or the bench the packing material is slowly poured through a funnel until no more can be accommodated no matter how long the tapping is continued. This may take a long time but care at this stage will be amply repaid later. It is convenient to pour the packing from a small beaker, weighing before and after, because this is a useful check on the packing density, particularly in metal columns. Metal columns are best packed when straight and then coiled. Sometimes vibration of the column is recommended but this tends to cause particle size separation unless very close particle size fractions are used.

Applications of gas-liquid chromatography

As already mentioned, gas-liquid chromatography is applied principally to the separation of volatile organic mixtures such as hydrocarbons, esters, alcohols, essential oils and many others. Additionally, many involatile or intractable compounds may be chromatographed after the formation of volatile derivatives. For example, sulphonic acids give rise to volatile compounds by chlorination or esterification[64], long chain fatty aldehydes by conversion to dimethyl acetals[65], carboxylic acids by esterification with diazomethane, sugars by methylation[66], amino-acids by conversion to *N*-trimethylsilyl[67] or *N*-trifluoroacetyl methyl esters[68], steroids to bromomethyl dimethylsilyl ethers[69] and glycerides and polyglycols by trimethylsilylation[70]. Fatty acids, phenols or barbituric acids in organic solvents have been refluxed with an alkylating agent and solid potassium carbonate, and the reaction mixture has been injected direct into the chromatograph, thus simplifying the often tedious derivatisation procedure[71]. Numerous other examples could be quoted; pyrolysis gas chromatography is used for the identification of polymeric and other involatile materials that do not lend themselves to the production of volatile derivatives, and is described in more detail below.

A glance through the literature – particularly Gas and Liquid Chromatography Abstracts[1] – will reveal the very wide applicability of gas chromatography, either to original mixtures, or to their volatile derivatives, and this aspect will not be elaborated here because it would be impossible to do it justice. Mention may be made, however, of expanding areas of interest which include biomedical applications of gas chromatography for diagnostic purposes and studies of metabolic disorders, as well as its use in forensic science for the detection of central nervous system stimulant drugs and alcohol. Gas chromatography is particularly valuable in pollution studies since many pollutants such as organophosphorus compounds and chlorinated hydrocarbons occur in trace quantities in, for example, wild life.

Gas-liquid chromatography may also be used for physicochemical measurements such as the determination of activity coefficients, heats of mixing and other thermodynamic quantities, as well as in studies of homogeneous and heterogeneous catalysis.

Pyrolysis gas chromatography

In recent years there has been increasing use of pyrolysis gas chromatography (PGC) for the identification of involatile materials, principally polymers, but also including drugs and substances of biological origin. Useful reviews have been given by Perry, McKinney and McFadden, and Levy[72]. Essentially the method is to decompose thermally a small sample of the test material and sweep the pyrolysis products on to a gas chromatographic column, where they will be partially or completely separated and emerge to give a characteristic chromatogram, often known as a pyrogram. A typical polymer pyrogram is shown in Fig. 4.17.

The method is simple both in principle and in practice and possesses the advantages both of low cost and short time of operation; only very small samples are required. Identification may be by simply comparing the pyrogram of the unknown with that of a standard substance. This is the so-called fingerprinting technique, and little, if any attempt is made to identify peaks in the pyrogram. It is the most widely applied use of PGC.

When the standard and the unknown pyrogram are

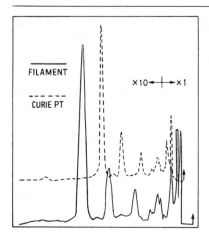

Fig. 4.17. Pyrogram of Solprene 1204

produced in the same laboratory only reasonable precautions are necessary to achieve acceptable reproducibility, particularly with materials whose nature is fairly well known, but when inter-laboratory comparisons are made very careful standardisation of conditions is essential. In order to compile a library of standard pyrograms the Pyrolysis Sub-group of the Chromatography Discussion Group has run a series of correlation trials[73] involving a number of laboratories. From the results it is hoped to be able to recommend standard conditions for the PGC of a number of polymeric materials. At the time of writing this has not proved possible, although the objective is probably in sight; the delay does highlight the difficulties of standardisation. Nevertheless pyrolysis gas chromatography is very widely used, but largely by laboratories working in isolation.

If no standard pyrogram is available for comparison with that of the unknown, further information about the test substance may be obtained only when care is taken to identify at least some of the peaks in the pyrogram by the usual techniques. From a knowledge of the nature of the pyrolysis products much useful information may be gleaned about the starting material and in the most favourable cases it may be sufficient for complete identification, perhaps with the aid of spectroscopic and other physical data. It may certainly be an aid in structure determinations and may provide useful confirmatory evidence in studies of thermal decompositions.

Experimentally, an ordinary gas chromatograph may be used with a pyrolysis unit in the carrier gas line, immediately before the column itself. This ensures that the pyrolytic fragments are swept rapidly and completely on to the column. For fingerprinting, practically any column that gives a reasonably detailed pyrogram will do, including capillary columns when only very small samples are available.

Pyrolysis units may be of several types but their common aim is to raise the temperature of the sample as quickly and as reproducibly as possible to ensure rapid and complete decomposition. Only a very thin layer of sample is used in order to avoid secondary reactions that may complicate the pyrogram. The most widely used pyrolysis units are of the filament type that are directly heated electrically, or Curie-point devices heated by induction, or microreactors.

Filament units

The sample is usually dissolved in a suitable solvent and coated in a thin layer on the filament. Passage of a small current suffices to evaporate the solvent. Catalytic effects do not appear to be important. After insertion of the filament into the chromatograph sufficient current is passed to raise the temperature to 500–700°C in a short time. One of the main disadvantages of the directly heated filament is that the time taken for it to reach its maximum temperature may be long enough for pyrolysis of the sample to occur at a temperature below that at which it is desired. The possibility thus arises of obtaining different sets of pyrolysis products from the same sample. Times of about 10 seconds have been recorded for the filament to reach its final temperature, although techniques are available for reducing this time to about 1 second[74].

Curie-point pyrolysers

Ferromagnetic conductors may be heated by induction to a constant known temperature in 20–30 milliseconds[75]. When a high frequency induction field is applied, the ferromagnetic material heats very rapidly up to the Curie temperature at which the absorption of energy decreases as the ferromagnetism disappears and, in effect, limits the temperature attained to this value. By choosing suitable alloys, a range of pyrolysis temperatures may be achieved, for example, pure iron has a Curie temperature of 770°C, an iron-nickel alloy 40% Fe/60% Ni 590°C, pure nickel 358°C.

The advantages of the Curie-point device are fairly obvious; a less obvious one is that extremely small samples may be conveniently handled. Pyrolysers of this type are available commercially.

Microreactors

Microreactors are usually small furnaces in which a constant hot zone is produced in the carrier gas line or sample loop. The sample is rapidly introduced into the hot zone either in a small boat, or perhaps by direct injection. The furnace may simply be a small quartz tube heated by an electric coil and suitably lagged. In one design use is made of a quartz tube packed with glass beads and inserted in a heated metal block.

The main advantage of the microreactor is that the pyrolysis temperature is accurately known and is constant; quantitative measurements are simple and any form of sample – gaseous, liquid or solid – may be pyrolysed.

Inorganic separations

Ellis and Iveson[76] have analysed quantitatively mixtures containing Cl_2, ClF, ClF_3, HF and UF_6. Naturally, special techniques are necessary in order to manipulate such corrosive gases and the apparatus was largely constructed of nickel and PTFE. The carrier gas was helium and one column packing was Kel-F oil supported on granulated PTFE. Phillips and Owens[77] have also tried PTFE capillaries coated with Kel-F oil for similar separations. Chlorides of tin and titanium, and niobium and tantalum have been separated by Keller and Freiser[78] using a column packed with Chromosorb coated with squalane or n-octadecane. The carrier gas was helium, flowrate about 50 cm³ per minute, and column temperature 100 to 200°C. Brandt[79] has separated the acetyl acetonates of beryllium and aluminium on a silicone oil column operating at 210°C. Bierman and Gesser[80] have also separated the acetyl acetonates of beryllium, aluminium, and chromium on a column using Apiezon L as the stationary phase.

Keulemans[81] has reported work in which certain inorganic substances were converted to their chlorides by grinding with carbon and chlorinating at 1000°C. Separations were then carried out on columns of molten anhydrous aluminium chloride supported on carbon, at temperatures up to 350°C. Volatile phosphorous compounds such as phosphorous trichloride, thiophosphoryl chloride ($PSCl_3$), phosphorous oxychloride, dimethyl and diethyl phosphite $(RO)_2POH$, have been separated by Shipotofsky and Moser[82], who used such liquids as Kel-F 90 and dibutyl phthalate supported on Fluoropak 80; column temperatures between 72 and 107°C, and carrier gas helium.

Organo-metallic compounds such as the tetramethyl derivatives of silicon, germanium, tin and lead have been separated by Abel, Nickless and Pollard[83] using an eight-foot column packed with 20% Apiezon L as the stationary phase, temperature $80°C$ and nitrogen carrier gas flowing at 50 cm³ per minute. Plots of the relative molecular mass of these compounds (and boiling points) against the logarithm of the retention times (see p. 211) are linear, similar to those of the hydrocarbons.

An example of the use of molten salt mixtures as stationary phases (see p. 224) is given by Juvet and Wachi[84] who have employed the bismuth trichloride-lead dichloride eutectic (m.p. $217°C$) to separate substances like titanium tetrachloride and antimony trichloride. Separation of the boranes and dimethyl boranes has been achieved by Seely, Oliver and Ritter[85] with a column packed with mineral oil on firebrick and helium carrier gas. Diborane, tetraborane and pentaborane have been completely separated by Kaufman, Todd and Koski[86]. Recently Schomburg[87] has discussed the use of gas-liquid chromatography in the chemistry of the boron alkyls and hydrides.

Cobalt has been determined in sub-nanogram amounts using 1,1,1,2,2,3,3-heptafluoro 7,7,-dimethyl-4,6,-octanediol, or H(fod)[88], to form volatile complexes which are separated by gas-liquid chromatography. The metal chelate possesses strong electron-capturing properties and the lower detection limit, using an electron-capture detector, is about 4×10^{-11} g cobalt. In a similar way beryllium[89] has been estimated in lunar dust and rocks by solvent extraction with trifluoroacetone (tfa) and gas chromatography of the $Be(tfa)_2$ complex. $Cr(tfa)_3$, prepared by heating the sample in the presence of tfa, may then be separated by chromatography. Zirconium, hafnium, thorium and uranium[90] apparently also form H(fod) chelates and hence their separation and determination are possible by methods similar to those used for cobalt.

Arsenic in biological materials has been determined as triphenylarsine[91] by incinerating the dried sample in a Schöniger flask, dissolving the residue in hydrochloric acid and extracting the arsenic as the diethylidithiocarbamate. This compound gives, with diphenyl magnesium, a quantitative yield of triphenylarsine which may be separated on a 5% Carbowax 20M column treated with terephthalic acid — temperature 220°C, carrier gas nitrogen.

Traces of hydrogen sulphide and sulphur dioxide[92] have been separated and estimated by means of a chromatograph in which contact with metal surfaces is minimised by maximum use of teflon; detection is by flame photometry. The support material is also teflon and the stationary phase Carbowax 400 or dinonyl phthalate. β-diketonates of aluminium, chromium and iron — principally of trifluoroacetyl-pivaloylmethane — have been successfully separated[93], with Apiezon greases as stationary phases, on a preparative scale.

It will be apparent from the above examples that to employ gas chromatography successfully for the determination of metals it is necessary to find a sufficiently volatile compound that is thermally stable at the column temperature employed and that may be prepared and extracted quantitatively from the sample. Various chelates have usually proved to be the most suitable derivatives. Once again, for further information reference may be made to Gas and Liquid Chromatography Abstracts[1] in which applications of gas chromatography to the separation of inorganic compounds have been comprehensively indexed.

Capillary column chromatography

The theory and applications of capillary columns were first developed by Golay[8, 94]. The theory predicts a very high efficiency for a column which is simply a long capillary whose inner wall is coated with a thin layer of stationary phase, and this prediction has been borne out in practice.

Modifications to the apparatus shown in Fig. 4.1 are

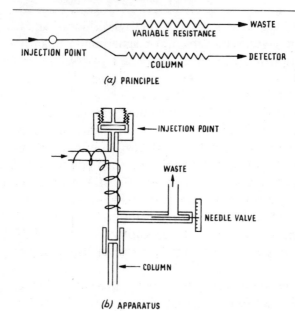

Fig. 4.18. Stream-splitter

necessary. They involve a special sample introduction technique, the capillary column itself and the use of a very sensitive detector of small volume and rapid response.

Sample introduction

When a capillary column is used sample sizes must be very small (less than 1 μg) because of the very small amount of stationary phase which can be held on the column walls. The simplest way of ensuring a reproducible sample which is at the same time small enough for efficient operation of the column is to use a 'stream splitting' device as shown schematically in Fig. 4.18 (*a*). The method is to inject a small sample of the mixture into the carrier gas stream in the usual way by means of a micro-syringe; the stream is divided so that only a known fraction of the carrier gas plus sample (say, 1/100) finds its way on to the column, the rest going to waste. It is desirable to heat the injection point. The splitting of the stream must be linear. That is, the proportion of each component of the mixture reaching the column must be the same as the proportion of carrier gas which flows through the column, regardless of changes in temperature, flowrate, sample size, 'split' ratio, and so on. An apparatus similar to that shown in Fig. 4.18 (*b*) has been thoroughly tested by Ettre and Averill[95] and shown to fulfil the requirements. The ratio of the 'split' is readily obtained by comparing the flowrate of the waste gas with that emerging from the column. In place of the needle-valve in Fig. 4.18 (*b*) a length of capillary can be used; changes in the 'split' ratio may then be obtained by varying the length or diameter of the capillary.

Flowrate through the column

The flowrate of the waste gas may be measured in the usual way, but the flow through the capillary is not easily determined by a soap-bubble flowmeter because although the linear velocity of the gas through the column is high, the volume which actually passes is small. Again, if a flame-ionisation detector is in use it is not possible to measure the flow of gas emerging from the column during an actual separation. The difficulties may be overcome by a preliminary calibration of the volume of gas (V) emerging from the column against $(p_i^2 - p_0^2)/p_0$, where p_i is the column inlet pressure and p_0 the outlet pressure (usually atmospheric). The volume of gas delivered at different inlet pressures may be measured by collection over water. A calibration should be carried out for each column temperature to be used. A ready estimation of the flowrate can thus be made from the column inlet pressure.

The column

Columns vary from 5 m to 300 m or more in length, with internal diameter of 0.25 mm to 1.25 mm. Materials commonly used are stainless steel, copper, nylon and glass. Capillaries in the first three materials at least are obtainable commercially. Glass capillaries may be constructed on a machine described by Desty, Haresnape and Whyman[96].

Column preparation

A number of methods have been described for coating the inner wall of capillaries with the stationary phase[77, 97] but that mentioned by Scott[98] appears to be the most satisfactory. About 1% of the column length is filled with a 10% solution of the stationary phase in a volatile solvent such as ether, and this volume of liquid is forced through the column at a speed of $2-5$ mm s^{-1} by gas; this rate is probably critical. Stronger solutions produce thicker films of the stationary phase. The solvent is removed by continuing the flow of gas through the column at a rate of about 1 cm^3 min^{-1}, for about an hour.

Glass capillaries appear to present some difficulties since non-polar stationary phases do not readily adhere to the walls, and the film breaks down into small lenses of liquid. In such cases thorough drying by forcing dry gas through the heated capillary (200°C), before coating is attempted, helps considerably. The liquid film on a capillary wall is analogous to a cylindrical soap-film. It is known that the soap film is unstable if its length is greater than about three times its diameter. It is to be expected, therefore, that films of stationary phases will be unstable unless there is moderately strong adhesion to the capillary wall.

The detector

The requirements of the detector have already been mentioned, namely, high sensitivity, small volume and rapid response. These three criteria are satisfied by most ionisation-type detectors.

Recorder

The time required for certain separations has been shortened so much that ordinary potentiometric recorders are unable to keep pace with the signals from the detector, so quickly do the separated components emerge from the column. In such cases other apparatus such as a cathode-ray oscilloscope[99] may be used to present the signal. Although such great rapidity is, perhaps, an extreme case, it is nevertheless advisable to employ a recorder with as fast a response as possible.

Scope of capillary columns

When fully exploited, capillary columns may well represent the ultimate in resolving power by chromatographic methods. Although this stage has not yet been reached it is clear that capillary columns can be more efficient than ordinary packed ones – Condon[97] says that they give up to five times the resolution in less time (see also Scott[98, 99] and Desty and Goldup[100]. The better resolution makes it possible to use carrier-gas flowrates above the optimum value (see p. 269), and still obtain separations comparable with those on packed columns. This can result in a marked decrease in the separation time.

It has been stated that capillary columns can separate mixtures while operating at lower temperatures than would normally be possible with a packed column, because of the extremely small amount of sample required. The use of very sensitive detectors also enables conventional columns to be operated at lower temperatures than normal since very small samples may also be used.

The disadvantages of using a capillary column are mainly those associated with the refinement of any technique, namely, the greater care required in operation. It is, for example, difficult to introduce the very small samples on to the column in a reproducible manner, and almost impossible to collect the separated components of a mixture for identification by independent means although, as mentioned above (p. 213) the total effluent from a capillary column may, under the right conditions, be fed directly into a mass spectrometer.

It seems probable that capillary columns will never replace packed columns in gas-liquid chromatography for many types of separation. Their main function would seem to be the performance of the more difficult separations or the speeding up of the more lengthy routine separations already performed on packed columns.

Gas-solid chromatography

As already mentioned at the beginning of the chapter, the practical application of gas-solid chromatography antedated that of gas-liquid, principally through the efforts of workers such as Turner[101], Claesson[3], Phillips[4] and Cremer[6] mainly in the decade 1942–1952. The techniques used were, however, principally those of frontal analysis and displacement development which are of much more limited application than elution analysis. Consequently it is not surprising that when Martin and James showed the practical feasibility of gas-liquid chromatography using the elution technique, the method was much more enthusiastically taken up than gas-solid chromatography. Apart, that is, from workers such as Ray[102] who used active carbon for the separation of permanent gases and low-boiling hydrocarbons; Janak[103] who separated nitric oxide, nitrous oxide, carbon monoxide and krypton on charcoal and low-boiling hydro-carbons on silica gel and on zeolites; McKenna and Idel-man[104]; Greene and Pust[105], Szulczweski and Higuchi[106] and Marvillet and Tranchant[107] who all worked with silica gel or alumina or both. Nevertheless, over the years, gas-solid chromatography has established itself as a powerful complementary technique to gas-liquid chromatog-raphy, mainly because it has been found possible to use the elution technique successfully, sometimes after modification of the surfaces of the adsorbents, as will be described below.

Technique of gas-solid chromatography

For elution analysis the same apparatus and technique are used as for gas-liquid chromatography. The only differences are the nature of the stationary phase – the column pack-ing – and the length of column required, hence in this account the column packing will be dealt with almost exclusively. Before doing so, however, it will be useful to look at the advantages and disadvantages of gas-solid as compared with gas-liquid chromatography. Active solids that will be described will include: carbon, alumina, silica gel, molecular sieves and porous polymers.

Solids are usually more thermally stable than liquids and possess lower vapour pressures, hence columns packed with active solids are not subject to bleeding at high temperatures, which upsets the detection system, quite apart from pro-ducing variations in retention volumes. Prolonged high temperature treatment of some active solids may result in variations in their surface properties, perhaps by loss of chemisorbed species, or by sintering (the reduction of the surface area at temperatures below the melting point), but such changes are unusual at the temperatures used in chromatography, or may be prevented by appropriate pre-treatment.

A column packed with an active solid may produce retention times greatly in excess of those obtained with a column of equal volume in gas-liquid chromatography, hence the more volatile substances can be separated conveniently — the molecular mass limit in gas-solid work is in fact about 150 (clearly this may also be a disadvantage.). Thus gas-solid chromatography is used mainly for separations of permanent gases and low-boiling materials.

If frontal analysis or displacement development is desirable gas-solid chromatography, on account of the sharp front boundaries that are produced, is far more effective than gas-liquid chromatography, although these techniques[4, 108] are used relatively infrequently.

The most obvious disadvantage of solid adsorbents is that they frequently give rise at low concentrations of adsorbate, to strongly curved isotherms, of the Langmuir type — see Curve IV, Fig. 4.20. As explained in Chapter 1, this leads to tailing with consequent loss of efficiency in elution analysis. Although gas-liquid systems also give curved isotherms (Curves I, II and III, Fig. 4.20) these are anti-Langmuir and approximately linear up to quite high concentrations. It may be noted that in plotting the isotherms in this book the convention for gas-solid systems is used, that is, the weight of vapour sorbed per gram of stationary phase — solid or liquid — is plotted against the relative pressure. This is to enable comparisons to be made between solid and liquid stationary phases; normally when discussing solutions of gases in liquids the Raoult Law plot of relative pressure (p/p^0) against mole fraction is employed.

When concentrations of sorbates are sufficiently high in gas-liquid chromatography for the curvature of the isotherm to become important, fronting occurs (Fig. 4.13 (*b*)), otherwise symmetrical peaks are the rule. In any case, bandspreading due to anti-Langmuir sorption behaviour is seldom as severe as in the converse circumstances, since the maximum concentration within a chromatographic band decreases steadily with distance down a column, and hence there is a tendency for the effective concentration range to be reduced to the linear part of the isotherm. With Langmuir isotherms very low concentrations, or very low relative pressures in the case of gases, are usually necessary before the isotherm is effectively linear.

As a result of peak asymmetry caused by curved isotherms, retention volumes measured to the peak maxima (Fig. 4.13) vary with sample size, although there is a reasonably constant point on each type of asymmetric chromatogram, namely, the end of the tail in tailing and the beginning of the front in fronting. Measurements made to these points are usually unsatisfactory, particularly with tailing since it is extremely difficult to locate the end of the tail.

Langmuir-type isotherms are the results of two properties of the adsorbent; microporosity and polarity. Polarity will give rise to such effects as hydrogen bonding in extreme cases, and under normal circumstances will enhance the normal van der Waals' attraction between the adsorbent and adsorbate, particularly if the latter is itself polar. By microporosity is meant the presence of pores with diameters up to about 5 nm (50 Å). Their presence does not, of course, preclude the presence of larger pores, that is, transitional pores, 5–50 nm in diameter and macropores with diameters in excess of 50 nm, but these will only affect the shape of the isotherm at fairly high relative pressures or concentrations of adsorbate and hence from the chromatographic point of view are usually unimportant. Strongly curved isotherms in the Langmuir sense imply high heats of adsorption and hence long retention times, except with very volatile materials.

Another problem with active solids is that, by their very nature, they are difficult to prepare with consistent properties from batch to batch, so that retention data may have to be continually checked if the column packing is frequently changed.

Finally, many active solids are efficient catalysts, whereas catalytic activity is rare in gas-liquid chromatography. Catalysis may result in the irreversible adsorption of a substance on the column, or the complete or partial conversion of substances giving rise to artefacts in the chromatogram. Loss of a component is, indeed, sometimes encountered even when it would seem most unlikely, for example, Phillips[109] has found that hexane appears to be adsorbed on to alumina by a fast and by a slow process, both of which are completely reversible. The effect of the slow process is so completely to attenuate the tail of the peak as to make part of it unobservable and hence unmeasurable. This effect could be caused by the presence of micropores or by lattice penetration, although the latter is usually confined to well crystallised materials such as zeolites.

Linear gas-solid chromatography

In order to use the elution technique successfully in gas-solid chromatography it is necessary to work with concentrations of adsorbates that correspond to the linear parts of their appropriate isotherms, or to try to straighten the isotherm. Complete linearity is unobtainable, but it is possible to get sufficiently close so that symmetrical elution peaks are obtained, and this after all is the principal objective. Also it is worth remembering that an anti-Langmuir isotherm is usually preferable to a Langmuir one; surface treatment of an active solid, silanisation, for example, will often change the isotherm for a particular adsorbate from Langmuir to anti-Langmuir.

There are three main methods for producing linearity in an isotherm, or for working on the most nearly linear portion of a curved isotherm:

Separation is effected at temperatures which are high relative to the boiling point of the sample being separated. As already mentioned, many separations by gas-solid chromatography are carried out on low-boiling materials.

Adsorbents with suitable surface characteristics are prepared. Examples are aluminas modified with inorganic salts; silicas modified by hydrothermal treatment and silanisation; graphitised carbon black; porous polymers.

The surface is pre-loaded with liquids to reduce surface inhomogeneity, for example, squalane on carbon black, or a carrier gas that interacts with the surface is used, such as water vapour with alumina or silica columns, or super-critical fluids are used.

Further examples of the application of these methods are given in the more detailed accounts of selected adsorbents which follow. A separate section (p. 257) deals with the application of supercritical fluids. The main points discussed above are summarised in Table 4.2.

Alumina

One of the most popular adsorbents for liquid chromatography (see Chapters 2 and 5), alumina has found quite wide application in gas-solid chromatography, although it does have some disadvantages. In the first place it is highly polar and therefore interacts strongly with polar molecules such as water. Secondly it possesses high catalytic activity and hence should be used with caution; for example, it may convert acetone to diacetone alcohol. Thirdly it is not readily prepared to yield a highly consistent product from the surface-chemical point of view.

Table 4.2   *Comparison of gas-liquid and gas-solid chromatography*

	G.L.C.	G.S.C.
1. Isotherm	Invariably anti-Langmuir tending to linear at low concentration of sorbate	Usually Langmuir, but may be anti-Langmuir depending on surface and surface treatment
2. Applicability	All volatile materials except the more permanent gases	Particularly useful for low boiling substances. Specific interactions may be exploited. Super-critical phase chromatography for higher molecular mass materials.
3. Thermal stability of stationary phase	Few stationary phases stable above 300°C	Usually good, but very active solids may age or sinter at high temperatures
4. Reproducibility of packing	Good	Good for less active materials, but more active substances tend to be variable.
5. Reactions on column	May occur due to interaction with the solid support or, very rarely, with the stationary liquid	Packing may catalyse some chemical change. Some physical desorption processes may be slow giving apparent loss of adsorbate.
6. Frontal analysis and displacement development	Not suitable	Suitable with solids giving Langmuir isotherm

Much of the early work on alumina was carried out by C. G. Scott[110] who concentrated on surface modified materials. He first worked with water deactivated samples, and found that the polarity was a minimum with approximately a monolayer of adsorbed water. Mixtures of hydrocarbons up to C_5 could then be separated effectively at room temperature without significant tailing, but a disadvantage was the need to pre-saturate the carrier gas to prevent water being stripped from the column.

Scott used BDH Chromatographic Grade alumina for liquid chromatography. He measured the polarity as a function of the retention of ethylene relative to non-polar ethane and propane. Activation by heating at temperatures up to 500°C increased the polarity presumably by loss of water. Conversely, adsorption of water reduced the polarity until a minimum was reached when the amount of water required to fill a monolayer had been adsorbed. Further water continued to reduce the activity – presumably by reducing the surface area – but increased the polarity.

Halasz[111] used a similar technique to separate low-boiling hydrocarbons – methane to 2-methylpropane – in about 30 seconds in a capillary, 2 m by 0.25 mm i.d., loosely packed with alumina 0.1 to 0.15 mm particle size. The carrier gas was nitrogen moistened over $Na_2SO_4 \cdot 10H_2O$.

Scott later extended his work on alumina to substances modified with sodium hydroxide and obtained results similar to those given by the water modified aluminas without, however, the need to presaturate the carrier gas. Static experiments showed that the isotherms for cyclohexane and benzene were substantially linear, but the one for benzene was slightly Langmuir in type and that for cyclohexane slightly anti-Langmuir. This also reflected the tendency to tailing and to fronting in these compounds. Aluminas were also modified with sodium chloride, bromide and iodide, and it was found that the retention of benzene relative to heptane increased markedly in the order:

$$OH^-; Cl^-; Br^-; I^-$$

A disadvantage of using the halide modified material was the long retention experienced, although an improvement was obtained when the alumina was sintered at 800°C before addition of modifier. Other interesting modifiers were CuCl, $AgNO_3$ and CdI_2. The first two gave very long retention times for alkenes, and advantage was taken of this in analysing for trace amounts of them. A very large sample of the alkene-containing material was passed through a column of

AgNO$_3$-modified alumina and this subtracted the alkene which could be subsequently displaced by carrier gas saturated with oct-1-ene.

Aluminas for gas-solid chromatography are supplied by Woelm. One of these is 'alkali treated' and has been used at 300°C for the separation of the normal hydrocarbons hexane to decane. Another, apparently untreated, alumina has been used for the separation of atmospheric gases and hydrocarbons from methane to pentane, including butenes.

R. P. W. Scott and Lawrence[112] observed that the retention volumes of alkanes on aluminas moderated by water, go through a maximum in the temperature range 25–150°C. This effect was even more marked with silica gel.

McCreery and Sawyer[113] have used aluminas and magnesium silicates modified with various salts, for example, alumina coated with 10% Na$_2$SO$_4$, Na$_2$MoO$_4$, NaCl, or Al$_2$(SO$_4$)$_3$. These aluminas were found to be useful for the separation of *cis*- and *trans*-isomers, chlorobenzene and the various dichlorbenzenes. For olefins the *cis*-isomer is retained more strongly than the *trans*-.

Kirkland has used a colloidal alumina known as 'Baymal' [a fibrillar boehmite, A1O(OH)] [114].

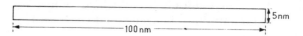

Fig. 4.19. 'Baymal' (Fibrillar Boehmite)

Baymal particles have been deposited on firebrick and on the walls of capillary columns. The specific surface of the boehmite is about 275 m^2 g^{-1}; it is non-porous, and it appears to be able to form adherent films from aqueous suspensions. Such films also have the property of anchoring colloidal silica onto surfaces, for example, glass beads may be dipped in a suspension of Baymal, washed, dried and then slurried with 3% silica sol. The silica adheres to the alumina which is in turn attached to the surfaces of the beads. The process may be repeated if necessary. The result is a so-called pellicular material, that is, composed of particles with impervious cores surrounded by porous layers — as described in Chapter 2 (p. 98). Columns made from this material have been used to separate natural gas, various unbranched hydrocarbon mixtures and aromatic hydrocarbons.

Genty and Schott[115] and West and Martson[116] have separated hydrogen isotopes on $Fe(OH)_3$ supported on alumina.

Little appears to have been published on the silanisation of alumina or the use of chemically bonded stationary phases as has been the case with silica.

Silica gel

It is probable that this adsorbent is somewhat easier to prepare in a reproducible form than is alumina — see Chapter 2 p. 30, and this may account for its more extensive use in gas-solid chromatography. In recent years commercial varieties of silica have become available that appear to offer distinct advantages over the usual gels which, when dry form glassy solids and have to be crushed and sieved to obtain appropriate particle sizes. The crushed particles are not uniform in shape and hence do not lend themselves to the uniform packing that is essential for an efficient column. Porasil (Pecheney-Saint-Gobain) is a spherical bead form of silica that is porous and can be supplied with pores that vary from 100 nm to about 150 nm in diameter. Because of the uniformity of the beads both from the point of view of the particle size and the porosity, they clearly offer advantages. The method of manufacture appears to involve the formation of an emulsion in a rather similar way to that involved in the manufacture of polystyrene beads. It is possible that the pore size is controlled by employing the so-called geometrical modification described later. Zipax, manufactured by du Pont is a related material in which the beads have impervious silica cores surrounded by a porous layer. Details of manufacture are not known but are possibly similar to the method described above for alumina using Baymal.

As would be expected, silica gel resembles alumina in its surface properties, that is, it is polar and catalytically active. As with alumina much work has been carried out to modify silica surfaces to render them suitable for gas-solid chromatography, not only by salts but also by chemical bonding of materials such as silanes and by hydrothermal treatment. Much of the early work on modified silicas was carried out by Kiselev and Scherbakova[117]. These workers refer to geometrical and to chemical modification of the surface.

As with most active solids silica gel tends to contain micropores and, as already mentioned, it is these pores, along with the polar surface, that give rise to strongly curved Langmuir isotherms. In particular the micropores are solely responsible for this behaviour with non-polar, non-polarisable adsorbates. Geometrical modification simply means the removal of the micropores by hydrothermal treatment involving contact with steam at 850°C for about 24 hours. The pore enlargement that results can be controlled by varying the time and temperature of treatment. Pores as large as 5000 nm can be generated. As expected, non-polar molecules, which give rise to symmetrical peaks, can be separated on such gels. Polar and polarisable molecules such as benzene, on the other hand, still give rise to tailing, mainly on account of surface hydroxyl groups. Tailing can be reduced by chemical modification.

As described by Kiselev and Scherbakova, chemical modification means silanisation of the surface hydroxyl groups with chlorotrimethyl- or dichlorodimethylsilane. Static measurements with benzene show a progressive change in the isotherm from Langmuir to anti-Langmuir as the degree of silanisation is increased. It is probably wise to carry out silanisation at an elevated temperature since there appears to be evidence that chlorosilanes will react readily with water at room temperature but not with surface hydroxyl groups until a temperature of about 200°C has been reached. Hence evolution of HCl at room temperature is no evidence of silanisation since reaction may be occurring with adsorbed water only. Nevertheless, some benefit may still be obtained since a silicone-like layer may cover some of the surface.

Silanisation renders the surface non-polar or nonspecific. Substances like ether and acetone that are normally very strongly adsorbed by silica gel are quickly eluted from a silanised column. As mentioned under solid supports, the process was originally developed to minimise adsorption effects on the siliceous solid supports used in gas-liquid work, but its success in gas-solid chromatography has led to the investigation of the use of other compounds in place of the chlorosilanes; that there is the possibility of esterifying the surface has been known for some time and both methanol and ethanol have been successfully bonded to the surface.

There have been two main approaches to the problem; first the formation of a chemically bonded stationary phase in proportions comparable to those in gas-liquid systems, and second, the formation of a monolayer of chemically bonded material as in silanisation.

The first approach is exemplified by early work carried out by Nickless and co-workers at Bristol, and later by Waddington and co-workers at York. For example, silanes have been polymerised in the presence of silica in the hope of obtaining a material chemically bonded to the silica surface but relatively little grafting of the silane polymer appears to occur. The co-hydrolysis of $SiCl_4$ with organic chlorosilanes is successful for the preparation of thermally stable – up to 350° – solids that are satisfactory for chromatography, but these solids are not coated on silica as in the previous example. Zipax materials used by Kirkland and De-Stefano[118] have also been prepared with chemically bound polymeric stationary phases, and it is claimed that by varying the functional groups the polarity can be varied, as may the film thickness and structure, that is, linear or cross-linked.

Using the second approach, Simpson has modified silica surfaces with alcohols in the series octanol up to the C_{22} compound. Other substances successfully used were Carbowax 20M, Ucon fluid HB 2000, β,β'-oxydipropionitrile and polyethylene glycol adipate (PEGA). Stable bonding seems to result from refluxing with the silica at 200–250°C under nitrogen in a Dean and Stark apparatus. In certain cases, for example, with benzyl alcohol, it is important to remove adsorbed water from the silica beforehand, otherwise gums and other polymers appear, as happens with the chlorosilanes as previously mentioned. All kinds of organic compounds are separable on the modified materials, although, once again, higher temperatures are required than would be needed in gas-liquid chromatography.

Porasil has also been modified by chemically bonding various materials to the surface, some of which are familiar as liquid stationary phases, such as Carbowax 400. Such modified beads go under the trade name Durapak stationary phases, the surface layer of modifier being limited to a monolayer. Little and co-workers[119] found that the polarity appeared to increase with increasing chain length of Carbowax, for example, from Carbowax 200 to Carbowax 2000, the opposite to what appears to happen when these materials are conventionally supported. Carbowax 400 bleeds at 40°C normally, but the Durapak material is stable up to 150°C.

Advantages claimed for such modified materials, apart from the enhanced thermal stability as far as the liquid phase is concerned, and the improved surface properties as far as the solid is concerned, are that the HETP (see p. 264) is independent of the nature of the sample, the temperature of the analysis and the sample size, and that mass transfer problems (see p. 269) are minimised.

Just as alumina has been modified by certain inorganic salts, so attempts have been made with the modification of siliceous materials. For example, Celite has been modified with the chlorides of lithium, sodium, potassium, rubidium and caesium, and these materials have been found effective for the separation of polycyclic aromatics. Proportions of salt used varied from 10%–20% by weight. Only slight variations in retention time occur in changing from one salt to another.

Hawton and Campbell[120] used Chromosorb P with 10% potassium tetraborate for the separation and identification of hydrogenated polyphenyls, using temperature programming from room temperature to 400°.

Carbon

Active carbon was the first adsorbent to be used for gas-solid chromatography, first by Claesson[3] and later by Phillips[4]. As already mentioned, these workers employed the displacement development technique. Ray[102] in 1954 used active carbon for the separation of the lower boiling hydrocarbon gases in preference to silica gel.

Active carbons are, perhaps, the most difficult solids to prepare in a reproducible form. They often have very high (nominal) surface areas and are usually microporous. One advantage they do possess, however, is that they are not very polar and often produce fronting peaks with water, indicative of an anti-Langmuir isotherm. The surface of normal active carbon or carbon black may sometimes be slightly polar owing to the presence of chemisorbed oxygen, sulphur or nitrogen. Eggertsen, Knight and Groennings[121] found that Pelletex carbon modified with 1.5% squalane gave improved peak symmetry, presumably by blocking of the polar parts of the surface.

The largely non-polar nature of the carbon surface is

useful, although the microporosity of active carbons can still lead to strongly curved isotherms and hence tailing. Graphitisation overcomes both polarity and microporosity problems and gives carbon with a high degree of surface homogeneity. Petov and Scherbakova[122] have used carbon blacks – surface areas 6–60 m^2 g^{-1} – graphitised at 3000°C introduced into the pores of hydrothermal silica by shaking for several hours. Such blacks usually have too small a particle size for column packing on their own. Columns made up with the silica-supported carbon gave symmetrical peaks with polar and non-polar substances, and with terpenes separation depended on the geometry of the molecule and its orientation on the basal faces of the graphite and not on the presence of functional groups or dipoles or on molecular masses. Guiochon[123] has modified the surface of graphite with phthalocyanins to reduce the elution time. With a surface layer of such molecules the carbon atomic density is reduced as compared with the graphite itself and hence adsorption energies are smaller, and compounds can be eluted at temperatures up to 100°C lower as a result, although the temperature limit of such a column is about 350°C. The inner walls of capillaries have been coated with graphite by a technique similar to that used for gas-liquid capillary columns, that is, by evaporation of the solvent from a graphite suspension[123].

Graphitised carbon blacks have been used by Liberti and co-workers[124] in packed capillaries. Columns are made by loosely packing a 2.2 mm i.d. glass tube with carbon black, down the centre of which a steel wire is inserted. The tubing is heated and drawn out to capillary dimensions such that the space left after withdrawal of the wire is 0.2–0.3 mm. The thickness of the carbon layer varies between 0.0045 and 0.0100 mm. It is claimed that such columns are non-specific, have high permeability, small pressure drop across them and high selectivity for certain systems. Geometrical and structural isomers such as *ortho-*, *meta-* and *para-*cresols and xylenes and polar compounds such as alcohols are examples, as well as isotopic systems containing deuterium, for example, deutero-acetone and acetone and deutero-nitromethane and nitromethane.

Molecular sieves

The term molecular sieve is applied nowadays to a wide range of materials that may be employed for the separation of mixtures by virtue of differences in molecular shape and size of the components. The molecular mass range is 10^1 to 10^5 and greater. In gas chromatography the sieves of importance are those that will separate materials at the lower end of this scale. Most sieves are of the zeolite kind that were thoroughly studied by Barrer. They consist of an aluminosilicate skeleton and contain pores of extremely regular size and shape formed by the stacking of polyhedra made up from SiO_4 and AlO_4 tetrahedra. Since the large internal surfaces of these solids derive from the nature of the crystal lattice rather than from random cracks and holes as in other active solids, their behaviour is much more predictable and consistent.

Table 4.3 describes three well known sieves (available commercially) and the properties associated with them; the effective pore diameter may be varied by ion exchange. As is well known, these sieves have a very high affinity for water, in fact, water will be adsorbed in preference to practically any other molecule, and this is why molecular sieves are used for drying carrier gases. Since the crystallite size is rather small it is usual to pelletise the material for convenience of handling. For gas-solid chromatography the pelletised material needs to be ground and sieved to provide fractions such as 40–60 mesh, or 60–80 mesh;

Molecular sieves are probably the most popular stationary phases for the separation of permanent gases, although, like

Table 4.3   *Properties of some molecular sieves*

	4A	5A	13X
Formula	$0.96\,Na_2O \cdot Al_2O_3.$ $1.92\,SiO_2 \cdot xH_2O$	Approx. 75% Na atoms in 4A replaced by Ca^{2+}.	$0.85\,Na_2O \cdot Al_2O_3.$ $2.49\,SiO_2 \cdot xH_2O$
Crystal structure	cubic	cubic	cubic
Pore diameter nm	0.42	0.42	up to 1.3
Effective pore diameter nm	0.40	0.50	less than 1.3
Specific surface $m^2\,g^{-1}$	700–800	700–800	700–800
External surface $m^2\,g^{-1}$	1–3	1–3	1–3

water, carbon dioxide is irreversibly adsorbed at room temperature. Hydrogen, nitrogen and oxygen are easily separated on 5A sieve at temperatures up to 100°C. Argon may be separated from oxygen at −78°C, or at room temperature if the column is long enough. Ortho- and para-hydrogen can be separated on 13X at −195°C but there is partial conversion of ortho- to para-. In spite of the considerable microporosity of these solids, even at the lowest temperatures just mentioned, separations are still being carried out at relatively high temperatures compared with the boiling points of the gases, and hence the isotherms are still effectively linear.

Other separations depend on the complete or nearly complete exclusion of some molecules but not others, for example, ethane (0.4 nm diameter) is adsorbed by 4A sieve but butane (0.5 nm) is not. Type 5A sieve will take up butane and unbranched alkanes but not branched chain compounds or cyclic compounds like benzene (0.63 nm diameter). On the other hand, on 10X and 13X, benzene is adsorbed in preference to unbranched alkanes because of π-electron interaction. Analyses are therefore possible in which a small molecule such as methane may be extracted from a mixture in a small pre-column of sieve before passing the rest of the mixture onto another column. The amount of material extracted may be estimated from the areas of the chromatograms obtained before and after using the pre-column.

On account of their microporosity, and like other adsorbents, molecular sieves require higher temperatures than does gas-liquid chromatography for a given separation, and this is frequently a disadvantage particularly with high boiling materials because the decomposition temperature may be approached, but occasionally molecular sieves offer specific advantages. For example, Brunnock and Luke[125] have found that by using a column of 13X sieve at 400–500°C it is possible to separate the naphthenes from the paraffins in petroleum distillates according to their carbon number. On a 1 m column and with temperature programming between 175 and 400°C samples could be separated which were large enough for fractions to be taken for subsequent analysis on a capillary column.

Microporous polymers

Co-polymerisation of styrene with divinyl benzene yields a porous polymer, the pore size depending, in part at least, on the proportion of divinyl benzene – which also produces the cross-linking. Such a system is, in fact, the basic polymer of many ion-exchange resins, which have ionisable groups grafted onto the aromatic nuclei. Ion-exchangers are about 10% cross-linked, microporous polymers about 30–40%. Hollis[126] first used these substances, and Poropak (Waters Associates) is probably the best known commercial variety. Surface areas of these materials are about 50 m^2 g^{-1}, and

Table 4.4

Poropak type	Polymer	Temperature limit
P	Ethylvinylbenzene/styrene	$250°C$
Q	Ethylvinylbenzene	$250°C$
R		
S	Modified with polar monomers	$250°C$
T		

average pore diameter 7.5 nm. Chromosorb 102, a styrene-divinylbenzene copolymer has a nominal area of $300-400$ m² g⁻¹ and pores $8.5-9.5$ nm.

It is claimed for these beads that partitioning occurs with the polymer itself, that is, they behave like 'solid' liquid stationary phases, but they can be modified with genuine liquid stationary phases in the same way as any other solid support.

With Poropak P and Q, polar compounds are rapidly eluted, but with R, S and T they are retained much more strongly. Nitrogen, oxygen and argon in air have been separated in that order at $-78°C$, and it is interesting to note that the order of elution is the reverse of that with molecular sieves or alumina. Generally speaking, the elution order appears to be in the order of increasing relative molecular mass – see the section on gel chromatography p. 62.

Various workers recommend conditioning of porous polymers before use, for example, 12 h at $200°C$ for normal use and 10 days for trace analysis. Rapid oxidation occurs in air at $200°C$. As with most adsorbents for gas-solid chromatography, porous polymers can only be used in the analysis of substances of relatively low molecular mass. Heavier substances produce long retention times and tailing.

Russian workers[127] have investigated thermally stable porous polymers based on polyimides. These materials possess large pore volumes and a large average pore radius – between 100 and 1000 nm. Surface areas are in the range 40 to 70 m² g⁻¹ , and the upper limit of temperature at which they may safely be used is in excess of 300°C. Owing to imide and carboxyl groups in the polymer surface, they give rise to specific molecular interactions, and thus interact strongly with polar adsorbates.

Supercritical phase chromatography

In the foregoing section on gas-solid chromatography it has been stressed that in general the technique is mainly suitable for the separation of substances of low molecular mass. It would be useful, however, if the advantages of adsorption chromatography — namely, the greater selectivity for substances of different chemical type — could be extended to substances of higher molecular mass. A method for doing this has been developed by Sie and Rijnders[128] by using high pressure gases, or supercritical fluids, instead of the usual relatively low pressure systems. Early work was done on alumina, but extension to porous polymers followed later.

In normal gas chromatography the carrier gas does not interact significantly with the stationary phase, but quite strong interactions occur if a vapour such as diethyl ether is used as the carrier gas at temperatures above its critical temperature. Under these conditions the adsorption equilibria of the substances being separated are shifted in a sense implying a much higher effective vapour pressure; that is, the substances appear to be much more volatile. Two factors mainly responsible for this are first, enhanced molecular interactions in the gas phase that increase the fugacity of the adsorbates, and, second, modifications to the adsorbent surface owing to enhanced interaction of the carrier gas. Pentane (T_{crit} 196.62°), diethyl ether (T_{crit} 193.61°), and propan-2-ol (T_{crit} 235.25°) are examples of substances that may be used as supercritical fluids at pressures within the range 40–60 kg cm⁻² .

Substances separated by this technique include vitamins A,

E_1 and K_1 and sex hormones. By ordinary gas-liquid chromatography these substances would have to be chemically modified or could not be chromatographed at all. By ordinary gas-solid chromatography they would decompose at the high temperatures required for volatilisation.

Organometallic substances such as ferrocene and some of its derivatives have also been separated successfully.

The theory of chromatography

Although it is possible to use chromatographic methods successfully without much knowledge of the principles, application of the theory is necessary for the fullest exploitation of the techniques, particularly in gas chromatography. Only the case of the linear isotherm will be dealt with here, hence the following is mainly relevant to partition chromatography; the isotherms in adsorption chromatography are almost always strongly curved. A theoretical treatment of curved isotherms has been given by Glueckauf[129], but it is rather complex. Of particular interest in gas-liquid chromatography are the 'plate' theory of Martin and Synge, subsequently modified by Glueckauf; the 'rate' theory of van Deemter *et al.* specially developed for gas-liquid chromatography; and the theory developed by Golay for capillary columns.

As indicated, the theories to be discussed assume linear isotherms, (see note on p. 240 on the convention adopted for isotherms) but in actual fact even 'partition' isotherms are often non-linear, particulary at high concentrations of the absorbate (Fig. 4.20); further, even if the system : vapour/stationary liquid forms an ideal solution (that is, obeys Raoult's Law), the corresponding partition isotherm is still curved – as may easily be verified – like those in the figure. The success of the theories depends on the fact that at the low concentrations encountered in gas-liquid chromatog-

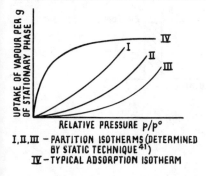

I, II, III – PARTITION ISOTHERMS (DETERMINED
BY STATIC TECHNIQUE[41])
IV – TYPICAL ADSORPTION ISOTHERM

Fig. 4.20. Isotherms

raphy the isotherms are effectively linear. As already mentioned (p. 240) partition isotherms curve in the opposite sense to the usual adsorption isotherm. The shape of the partition isotherms shown in Fig. 4.20 implies that chromatographic bands (or peaks) associated with these systems will possess diffuse fronts and sharp tails ('fronting') – Chapter 1. Such asymmetric peaks are often observed in gas-liquid chromatography particularly when large samples are used. Tailing is usually the result of adsorption on the solid support, adsorption at the gas/liquid interface, or to non-equilibrium effects. Under certain conditions, and with certain solute-solvent systems tailing may arise independently of these effects, but such situations are rare.

According to the simple theories of Wilson[130] and de Vault[131] a solute/solvent system giving a linear isotherm will also give chromatographic bands similar to those depicted in Fig. 4.21, that is, the initial band will travel unchanged on elution down the column. In fact, solutes do not behave like this, but as shown in Fig. 4.22, that is, they tend to form symmetrical (Gaussian) peaks which broaden during their passage down the column; the peak maximum at the same time becomes smaller. The 'plate' theory, originally developed by Martin and Synge[2], explains this behaviour with a fair degree of success.

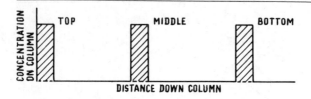

Fig. 4.21. Chromatographic bands (Wilson[130])

The plate theory

By analogy with distillation columns, chromatographic columns can be considered to be made up from a number of 'theoretical plates'. A theoretical plate is defined as a small section of the column in which equilibrium is established between the gas and dissolved phases of a solute before the gas phase passes on to the next stage, where the process is repeated. The concept of the 'plate' may be better illustrated by reference to the Craig 'machine' which is a liquid-liquid extraction device and bears a strong resemblance, in its mode of action, to liquid-liquid chromatography. It is essentially a large collection of separating funnels in each of which two immiscible solvents can be shaken together, allowed to separate, and the upper layer of solvent passed to the next funnel. Each funnel corresponds to a theoretical plate. Some idea can now be given how the peak shape arises in a partition column and why band broadening occurs.

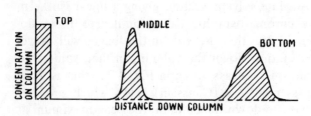

Fig. 4.22. Chromatographic bands—gas-liquid chromatography

If a substance A is introduced into the first stage of a Craig apparatus, and then shaken until equilibrium is established, with two immiscible solvents; M, the upper and mobile layer, and S, the lower and stationary layer, A will distribute itself between M and S according to its partition coefficient k. Solvent M corresponds to the mobile phase and S corresponds to the stationary phase in chromatography. Next, the upper layer M, containing a proportion of A, is transferred to the second stage and an equal quantity of fresh M is introduced into the first stage. The process is repeated by shaking both stages and then transferring the layers of M from the second to the third stage and the first to the second, fresh M again being introduced into the first stage. The process is repeated n times, during which the solvent layer M moves successively from stage to stage while S remains stationary. The distribution of A among the n stages by the process just described, is obtained by expanding the expression

$$(a + b)^{n - 1}$$

where n is the number of stages, $a = 1/(k + 1)$, $b = k/(k + 1)$, and k is the partition coefficient, defined as:

$$k = \frac{\text{concentration of solute in stationary phase } (S)}{\text{concentration of solute in mobile phase } (M)}$$

If the volumes V_m and V_s of the solvents M and S are unequal, k must be multiplied by V_m/V_s. Fig. 4.23 shows the distribution of A after three transfers, assuming equal volumes of M and S in each stage and partition coefficient $k_A = 1$.

The distribution of two substances, A and B, initially mixed and dissolved in the two solvents of the first stage, will next be considered. For simplicity it is assumed that the amounts of M and S in each stage are equal, and the partition coefficients of A and B are 2 and ½ respectively. The

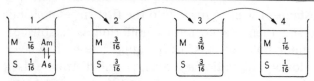

FRACTIONS REFER TO AMOUNT OF SUBSTANCE 'A' IN EACH LAYER

Fig. 4.23. Illustration of operation of Craig machine, situation after three transfers

distributions of A and B after nine equibrations, are obtained by expanding the expressions

$$(2/3 + 1/3)^8 \text{ for } A, \text{ and } (1/3 + 2/3)^8 \text{ for } B, \text{ namely}$$

No. of stage	1	2	3	4	5	6	7	8	9	
Fraction of A	$\dfrac{256}{6561}$	$\dfrac{1024}{6561}$	$\dfrac{1792}{6561}$	$\dfrac{1792}{6561}$	$\dfrac{1120}{6561}$	$\dfrac{348}{6561}$	$\dfrac{112}{6561}$	$\dfrac{16}{6561}$	$\dfrac{1}{6561}$	(Total = 1)
Fraction of B	$\dfrac{1}{6561}$	$\dfrac{16}{6561}$	$\dfrac{112}{6561}$	$\dfrac{348}{6561}$	$\dfrac{1120}{6561}$	$\dfrac{1792}{6561}$	$\dfrac{1792}{6561}$	$\dfrac{1024}{6561}$	$\dfrac{256}{6561}$	(Total = 1)

Distribution curves (which correspond to elution peaks) are plotted in Fig. 4.24 and from this simple treatment it may be seen that substances which possess different partition coefficients will have their maximum concentrations moving at different rates, but can never be *completely* separated, although after a sufficient number of stages the fractional band impurity (Glueckauf[132]) – see Fig. 4.24 – will be very small.

When n is large the amounts of the substance in the first and last few stages may be neglected. Distribution curves tend to be Gaussian for large values of n, and the larger the number of stages or plates the wider will be the elution peak. In chromatography it is possible to estimate the number of theoretical plates from the peak width, and the time taken to elute the peak maximum – see Fig. 4.13 (a) – namely:

$$n = 16 \frac{x^2}{y^2}$$

where n is the number of theoretical plates, x is the horizontal distance between the injection point and the component maximum, and y is the width of the component peak.

The plate number, n, is a measure of the efficiency of the column and may be increased simply by increasing the length of the column. A proportional increase in retention time (or volume) occurs, but the peak width increases and the peak height decreases as the square root of n. In order to increase efficiency it is therefore advantageous to increase the column length, but this process cannot be continued indefinitely since a penalty is paid in terms of increased analysis time. Further, a law of diminishing returns operates, since, as the column length increases so the ratio of the inlet pressure (p_i) to the outlet pressure (p_0) also increases and this tends to

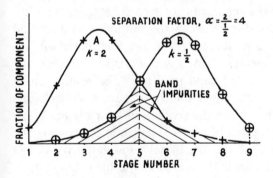

Fig. 4.24. Separation in a counter-current apparatus

reduce the efficiency as is shown later in the discussion on the rate theory.

Although, as stated, the plate number, n, is a measure of the efficiency of the column, for comparative purposes a more convenient quantity is the Height Equivalent to a Theoretical Plate (HETP) which is given by:

$$HETP = L/n$$

where L is the length of the column. It is seen that the smaller the HETP the better the column.

The simple theory illustrates several important features but its obvious limitation is that each stage of the Craig machine is fully equilibrated before the mobile phase is moved on to the next stage. In chromatography, dynamic systems involving continuous flow are being dealt with. Also, gases are compressible so that flowrates vary along the column; hence the theory must be modified to account for these factors. Glueckauf[132] has derived equations describing the distribution of a solute under the conditions of continuous flow. This distribution is 'Poisson' and like the binomial distribution described above approximates to 'Gaussian' when the plate number (n) is greater than about 100.

On considering the separation of two substances, Glueckauf derives a relationship between the number of theoretical plates, the separation factor (ratio of the partition co-efficient — always greater than 1, see 'Partition coefficients' below), and the fractional band impurity. The results can be represented conveniently as a family of curves. Glueckauf's treatment shows that when the separation factor is small, say 1.03, a very large number of plates is required to separate the two substances so that the fractional band impurity is less than 1% (more than 30,000), and that the band impurity is greatest when the mole fractions of the substances to be separated are equal. This means that *better* separation is achieved when the mole fractions are unequal. A particularly clear exposition of the continuous flow model has been given by Keulemans[133].

Peak resolution

For two substances, A and B, giving rise to Gaussian peaks, the peak resolution, R, may be defined as:

$$R = \frac{2 \, \Delta x}{y_A + y_B}$$

where Δx (Fig. 4.25) is the distance between the peak maxima and y_A and y_B are the peak widths at the bases.

The equation is also applicable to peaks which are completely separated.

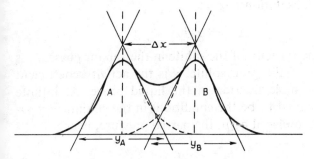

Fig. 4.25. Peak resolution

Partition coefficients

From the foregoing discussion it will be apparent that two substances A and B are separated by virtue of the difference in their partition coefficients, k_A and k_B. Comparing these partition coefficients such that:

$$\frac{k_A}{k_B} = \alpha; \quad \alpha \geqslant 1$$

where α is known as the separation factor (see above), it will be seen that if $\alpha = 1$ no separation can occur, and the larger the value of α the easier will be the separation. Partition coefficients can be estimated from chromatographic data since:

$$V_N = kV_L$$

where V_L is the volume of liquid on the column. Alternatively, one may use the relationship:

$$V_g = \frac{273 \cdot k}{T\rho}$$

where ρ is the density of the stationary phase and T the temperature of the column.

In non-ideal solutions:

$$f = \gamma x f_0$$

where f is the fugacity of the solute in the vapour phase, f_0 is the fugacity of the pure solute, γ is the activity coefficient and x is the mole fraction in the liquid phase. At infinite dilution, assumed to be the conditions in the column, $\gamma = \gamma^\infty$ and f may be replaced by p, the partial pressure, hence:

$$p = \gamma^\infty x f_0$$

using the definition of k quoted earlier, that is:

$$k = \frac{\text{concentration in liquid}}{\text{concentration in gas}}$$

then

$$k = \frac{N_{\text{liq}} \cdot RT}{\gamma^\infty f_0}$$

where N_{liq} is the number of moles of stationary phase per unit volume. It is now possible to write:

$$V_N = \frac{N_{liq} \cdot RT \cdot V_L}{\gamma^\infty f_0}$$

or

$$V_g = 273 \cdot \frac{R \cdot N_{liq}}{\gamma^\infty f_0}$$

and it is evident that the activity coefficient at infinite dilution is of fundamental importance, since:

$$\alpha_{AB} = \frac{k_A}{k_B} = \frac{(\gamma^\infty f_0)_B}{(\gamma^\infty f_0)_A}$$

Hence, for a given separation, if the activity coefficients of two substances A and B in the stationary liquid are similar, one must rely on the difference in vapour pressure. Conversely, if the vapour pressures are similar, one must find a liquid in which the activity coefficients (and hence the partition coefficients) differ sufficiently. An interesting example is given by benzene and cyclohexane, two substances whose boiling points differ by less than 1°C; cyclohexane is eluted first from a dinonyl phthalate column, but benzene is eluted first from a squalane column. This is entirely a reflection of the difference in the activity coefficients of benzene in the two solvents.

From the temperature variation of k it is possible to calculate heats of solution[10, 134] and provided that the linear isotherms are obtained it is possible to determine heats of adsorption in gas-solid chromatography by an analogous process[135]. A number of workers have determined γ^∞ values by gas-liquid chromatography[136–139] and Phillips[140] has determined the free energy, heat, and entropy of complex formation.

The rate theory

The rate theory[141] employs a kinetic approach to gas-liquid chromatography and explains band broadening in terms of a number of rate factors. The equation ('van Deemter equation') derived is:

$$\text{HETP} = 2\lambda d_p + \frac{2\gamma D_G}{\bar{u}} + \frac{8}{\pi^2}\frac{k'}{(1+k')^2}\frac{d_f^2\,\bar{u}}{D_L}$$

$$\underbrace{\underset{\substack{\text{eddy} \\ \text{diffusion}}}{} \quad \underset{\substack{\text{molecular} \\ \text{diffusion}}}{}}_{\text{axial diffusion}} \qquad \text{non-equilibrium effect}$$

(The symbols are defined at the end of this chapter)

The van Deemter equation may be written in the simplified form:

$$\text{HETP} = A + \frac{B}{\bar{u}} + C\bar{u}$$

where $\bar{u}$, the linear gas velocity (cm sec^{-1}), is obtained by dividing the column length by the time taken for the air-peak to traverse the column.

If HETP is plotted against $\bar{u}$ a curve (hyperbola) is obtained similar to that shown in Fig. 4.26. It will be seen that for maximum efficiency the constants A, B, and C should be as small as possible.

A is the 'eddy diffusion' term. Golay has suggested that the name 'multipath term' be used instead of eddy diffusion, and that would seem to be more appropriate since it arises from the different path lengths travelled by different molecules. It depends on the particle shape and size of the supporting material, and is therefore a characteristic of the column packing, and constant for a given column. In theory A can be reduced by reducing the particle size, d_p, but below

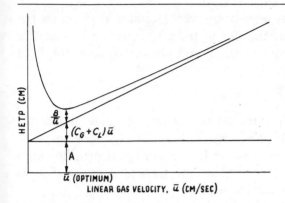

Fig. 4.26. Plot of van Deemter equation

a certain level there is an increase in γ, because of the difficulty of packing very fine powders. The best particle size seems to be within the range 50–80 B.S.S. The fairly coarse particles are also less likely to give rise to clogging of the column which reduces the permeability.

$B/\bar{u}$ refers to the molecular diffusion of the solute molecules, and it can be seen from the graph that for small values of $\bar{u}$ the term makes the largest contribution to HETP, while for large values of $\bar{u}$ the contribution becomes negligible. Since B is proportional to the diffusion coefficient of the carrier gas the heavier gases such as carbon dioxide tend to give the best performance at low flowrates, that is, smaller values of HETP are obtained.

$C\bar{u}$, the non-equilibrium effect, is due to the finite rate of mass transfer in the liquid phase, that is, the time taken for solute molecules to transfer from the stationary phase to the gas phase. This term varies in the opposite sense to $B/\bar{u}$, that is, its contribution to HETP increases with increasing flowrate, when it tends to become the most important of the three terms in the van Deemter equation. It is also affected by the thickness of the liquid film (d_f), although not in a straightforward manner since k' is also affected.

It has been suggested by van Deemter that, since the resistance to mass transfer in the gas phase is not negligible compared with that in the liquid phase, the constant, C, is better represented by a composite term:

$$C = C_G + C_L$$

where C_G and C_L refer to the resistance to mass transfer in the gas and liquid phases respectively. The lighter gases, such as hydrogen and helium, are better carriers at high flowrates because C_G is smaller than for heavier gases, and B is negligible even for light gases.

The value of the rate theory, which adopts a more realistic approach than the plate theory to the mechanism by which band spreading occurs in a chromatographic column, lies in its usefulness in suggesting ways in which the performance of a column can be improved[53, 142, 143]. Apart from the fairly obvious suggestion that increasing the length of the column will improve the efficiency, the plate theory is not very helpful in this respect. The van Deemter equation is applied by plotting the HETP – as determined from the peak width (p. 263) – against $\bar{u}$ for variations in such parameters as column size, particle size of solid support, amount of stationary phase, nature of carrier-gas and so on. Families of curves, similar to that shown in Fig. 4.26, are obtained, in each case with a minimum corresponding to the optimum flowrate of the carrier gas. The shape of the curve in Fig. 4.26 also suggests that it is better to employ too high a flowrate than too low.

The van Deemter equation also reveals the importance of the flowrate (or, more precisely, the linear velocity) of the carrier gas in determining the efficiency of the chromatographic process. The flowrate, and hence $\bar{u}$, depends on the pressure drop across the column ($p_i - p_0$), where p_i and p_0 are the column inlet and outlet pressures respectively. However, the measured value of $\bar{u}$ is an average for the whole length of the columns because of the compressibility of the gas. The largest variation in $\bar{u}$ occurs near the column outlet, and this variation increases with increase in the value of p_i/p_0. It is therefore an advantage to keep this ratio as near as possible to unity. It follows that in order to obtain a given pressure drop across the column it is better to increase p_i rather than to decrease p_0 since the value of p_i/p_0 will be nearer to unity in the former case, and hence less variation in $\bar{u}$ will occur.

Values of p_i/p_0 up to about 1.5 give an almost negligible variation in $\bar{u}$ along the column, so that if $\bar{u}$ happens to be the optimum linear gas velocity then most of the column will be operating at maximum efficiency.

Golay theory

As already mentioned, the Golay theory[94, 144] concerns itself with tubular columns in which the stationary phase is held on the walls of the tube only. Golay derives expressions for the HETP for round and rectangular columns. For a round capillary Golay's equation for determining the HETP may be written in the form (*cf.* van Deemter equation):

$$\text{HETP} = \frac{2D_G}{\bar{u}} + \left[\frac{1 + 6\,k' + 11(k')^2}{24(1 + k')^2} \cdot \frac{r^2}{D_G} + \frac{2}{3} \frac{k'}{(1 + k')^2} \frac{d_f^2}{D_L} \right] \cdot \bar{u}$$

which, once again, can be simplified to

$$\text{HETP} = \frac{B'}{\bar{u}} + C'_G \cdot \bar{u} + C'_L \cdot \bar{u}$$

and used in a similar way to the van Deemter equation. B has the same significance as before, and C_G' and C_L' account for the resistances to mass transfer in the gaseous and liquid phases respectively. The term involving the column packing (A) is not required. Golay also derives another expression which he calls the 'Performance Index' of a column. It measures the *overall* efficiency of the chromatographic column.

Symbols used in the van Deemter and Golay equations

HETP	height equivalent to a theoretical plate
λ	a measure of the packing irregularities
d_p	particle diameter
γ	tortuosity factor
D_G	coefficient of gaseous diffusion
D_L	coefficient of liquid diffusion
$\bar{u}$	average linear gas velocity
k	partition coefficient
k'	ratio of the capacity of the liquid phase to that of gas phase. For capillaries $k' = 2\,kd_f/r$
d_f	average film thickness
r	radius of capillary

REFERENCES

1. Knapman, C. E. H., *Gas Chromatography Abstracts*, 1958–1962 Butterworths, London; 1963–1969 Institute of Petroleum, London; *Gas and Liquid Chromatography Abstracts*, 1970–1972 Institute of Petroleum, London; 1973 *et seq.* Applied Science Publishers, Barking, England.
2. Martin, A. J. P., Synge, R. L. M., *Biochem. J.*, 1941, **35**, 1358.
3. Claesson, S., *Arkiv Kemi, Min. Geol.*, 1946, **A23**, 1.
4. Phillips, C. S. G., *Discuss. Faraday Soc.*, 1949, No. 7, 241. Griffiths, J., James, D., Phillips, C. S. G., *Analyst*, 1952, **77**, 897.
5. Glueckauf, E., Barker, K. H., Kitt, G. P., *Discuss. Faraday Soc.*, 1949, No. 7, 199.
6. Cremer, E., Prior, F., *Z. Electrochem.*, 1951, **55**, 66. Cremer, E., Muller, R., *ibid.*, p. 217.
7. Martin, A. J. P., James, A. T., *Biochem. J.*, 1952, **50**, 679.

8. Golay, M. J. E., *Gas Chromatography*, Editors Coates, Noebels and Fagerson, Academic Press, New York, 1959, p. 1.

9. Stack, M. V. *Gas Chromatography* 1968, Editor Harbourne, C. L. A. Institute of Petroleum, London 1969, p. 109.

10. Littlewood, A. B., Phillips, C. S. G., Price, D. T., *J. Chem. Soc.* 1955, 1480.

11. James, D., Phillips, C. S. G., *J. Sci. Inst.*, 1952, **29**, 362.

12. Hall, R. A., *ibid*, 1955, **32**, 116.

13. Harrison, G. F., *Vapour Phase Chromatography* 1956, Editor Desty, D. H., Butterworth, London, 1956, p. 332.

14. Scott, R. P. W., *Gas Chromatography* 1958, Editor Desty, D. H., Butterworth, London, 1958, p. 189.

15. McCreadie, S. W. S., Williams, A. F., *J. Appl. Chem.*, 1957, **7**, 47.

16. Otte, E., Jentzsch, D., *Gas Chromatography* 1970, Editor Stock, R., Institute of Petroleum, London, 1971, p. 218.

17. Kolb, B., Hoser, W., *Chromatographia* 1973, **6**, 28.

18. Astbury, G. K., Davies, A. J., Drinkwater, J. W., *Analyt. Chem.*, 1957, **29**, 918.

19. Fischer, H., Neufelder, M., Pruggmeyer, D. *Chromatographia* 1972, **8**, 613.

20. Halasz, I., *Analyt. Chem.* 1964, **36**, 1428.

21. Guiochon, G., Goedert, M., Jacob, L. *Gas Chromatography* 1970, Editor Stock, R., Institute of Petroleum, London 1971, p. 160.

22. Clarke, A., Grant, D. W. *ibid.*, p. 189; Wittebrood, R. T. *Chromatographia* 1972, **8**, 454.

23. Littlewood, A. B., *Gas Chromatography*, Academic Press New York, 1970 p. 339.

24. Martin, A. J. P., James, A. T., *Biochem. J.*, 1956, **63**, 138.

25. Phillips, C. S. G., Timms, P. L., *J. Chromatog.*, 1961, **5**, 131.

26. McWilliam, I. G., Dewar, R. A., *Gas Chromatography* 1958, Editor Desty, D. H. Butterworth, London, 1958, p. 142.

27. Lovelock, J. E., *J. Chromatog.*, 1958, **1**, 35; *Gas Chromatography* 1960, Editor Scott, R. P. W., Butterworth, London, 1960, p. 16.

28. Lovelock, J. E., *Analyt. Chem.*, 1961, **33**, 162.

29. Condon, R. D., Scholly, P. R., Averill, W., *Gas Chromatography 1960*, Editor Scott, R. P. W., Butterworth, London. 1960, p. 30.

30. Willis, V., *Nature*, 1959, **184**, 894.

31. Berry, R., *Nature*, 1960, **186**, 578, *Gas Chromatography* 1962, Editor van Swaay, M., Butterworth, London, 1962, p. 321.
32. Bevan, S. C., Thorburn, S., *J. Chromatog.* 1965, **11**, 301; *Chem. Brit.*, 1965, **2**, 206.
33. Janak, J., *Chem. Listy*, 1953, **47**, 464, *Collection Czech. Chem. Communs.*, 1954, **19**, 684.
34. Ambrose, D., Kuelemans, A. I. M., Purnell. J. H., *Analyt. Chem.*, 1958, **30**, 1582.
35. Preliminary recommendations on nomenclature and presentation of data in gas chromatography. *Pure and Applied Chemistry*, **1**, No. 1.
36. Thombs, D. A., *Chromatographia* 1973, **6**, 111.
37. Ambrose, D., Purnell, J. H., *Gas Chromatography* 1958, Editor Desty, D. H., Butterworth, London, 1958, p. 363.
38. Adlard, E. R.. Khan, M. A., Whitman, B. T., *Gas Chromatography* 1960, Editor Scott, R. P. W., Butterworth, London 1960, p. 251.
39. Kovats, E., *et al.*, *Helv. Chim. Acta*, 1958, **41**, 1915; *ibid.*, 1959, **42**, 2519; *ibid.*, 2709.
40. Smith, J. F., *Chem. and Ind.*, 1960, 1024. Evans, M. B., Smith, J. F., *J. Chromatog.*, 1961, **5**, 300; 1961 **6**, 293.
41. Peetre, I-B., *Chromatographia* 1973, **6**, 257.
42. Haarhoff, P. C., van der Linde, H. J. *Analyt. Chem.* 1966, **38**, 573; Sewell, P. A., Stock, R., *J. Chromatog.* 1970, **50**, 10.
43. Watson, J. T., Biemann, K., *Analyt. Chem.*, 1964, **36**, 1135; Ryhage, R., *ibid.* p. 759; Henneberg, D., Schomburg, G., *Gas Chromatography* 1970, Editor Stock, R., Institute of Petroleum, London, 1971, p. 141.
44. Neuner-Jehle, N., Etzweiler, F., Zarske, G., *Chromatographia*, 1973, **6**, 211.
45. Anthony, G. M., Brooks, C. J. W., *Gas Chromatography* 1970, Editor Stock, R., Institute of Petroleum, London 1971, p. 70.
46. Guiochon, G., Goedert, M., Jacob, L. *ibid.* p. 160.
47. Deans, D. R., *Chromatographia* 1968, **1**, 187.
48. Stock, R., Mainwaring, J., unpublished work.
49. Ettre, L. S., *J. Chromatog.*, 1960, **4**, 166.
50. Drew, C. M., Bens, E. M., *Gas Chromatography* 1968, Institute of Petroleum, London 1969, p. 3.
51. Bens, E. M., *Analyt. Chem.*, 1961, **33**, 179.
52. Knight, H. S., *ibid.*, 1958, **30**, 2030.

53. Littlewood, A. B., *Gas Chromatography* 1958, Editor Desty, D. H., Butterworth, London, 1968, p. 23.
54. Halasz, I., Wegner, E. E., *Nature*, 1961, **189**, 570.
55. Keulemans, A. I. M., Kwantes, A., Zaal, P., *Analyt. Chim. Acta*, 1955, **13**, 357.
56. *Materials for Gas Chromatography*, 2nd Edition. May and Baker (Dagenham, England).
57. Porter, R. S., Johnson, J. F., *Analyt. Chem.*, 1961, **33**, 1152.
58. Cartoni, G. P., Lowrie, R. S., Phillips, C. S. G., Venanzi, L. M., *Gas Chromatography* 1960, Editor Scott, R. P. W., Butterworth, London, 1960, p. 273; Maczek, A. O. S., Phillips, C. S. G., *ibid.*, p. 284.
59. Schurig, V., Chang, R. C., Zlatkis, A., *Chromatographia* 1973, **6**, 223.
60. Freeguard, G. F., Stock, R., *Gas Chromatography* 1962, Editor van Swaay, M., Butterworth, London, 1962, p. 102.
61. Spencer, C. F., Johnson, J. F., *Analyt. Chem.*, 1960, **32**, 1386.
62. Martin, R. L., *ibid.*, 1961, **33**, 347.
63. Watson, D., Cocker Chemical Co., Oswaldtwistle. Personal communication.
64. Kirkland, J. J., *Analyt. Chem.*, 1960, **32**, 1388.
65. Gray, G. M., *J. Chromatog.*, 1960, **4**, 52.
66. Kircher, H. W., *Analyt. Chem.*, 1960, **32**, 1103.
67. Rühlmann, O. K., Giesecke, W., *Angew Chem.* 1961, **73**, 113.
68. Bayer, E., *Gas Chromatography* 1958, Editor Desty, D. H., Butterworth, London 1958, p. 333.
69. Eaborn, C., Walton, D. R. M., Thomas, B. S., *Chem. Ind.* 1967, 827.
70. Kresze, G., Schauffelhaut, F., *Z. anal. Chem.* 1967, **229**, 401.
71. Dunges, W., *Chromatographia* 1973, **6**, 196.
72. Perry, S. G., *Adv. in Chromatog.* 1968, **7**, 221; McKinney, R. W., McFadden, W. H., *Ancillary Techniques of Gas Chromatography*, Editor Ettre, L. S., Wiley-Interscience, New York 1969, p. 55; Levy, R. L., *Chrom. Rev.* 1967, **8**, 49.
73. Coupe, N. B., Jones, C. E. R., Perry, S. G. *J. Chrom.* 1970, **47**, 291; and *Gas Chromatography* 1970, Editor Stock R., Institute of Petroleum London 1971, p. 399; Coupe, N. B., Jones, C. E. R., Stockwell, P. B., *Chromatographia* 1973, **6**, 483.
74. Lerhle, R. S., Robb, J. C., *J. Gas. Chrom.* 1967, **5**, 89.

75. Simon, W., Giacobbo, H., *Chem-Ingr-Techr.* 1965, **37**, 709.
76. Ellis, J. F., Iveson, G., *Gas Chromatography 1958*, Editor Desty, D. H. Butterworth, London, 1958, p. 300.
77. Phillips, T. R., Owens, D. R., *Gas Chromatography 1960*, Editor Scott, R. P. W., Butterworth, London, 1960, p. 308.
78. Keller, R. A. Freiser, H., *ibid.*, p. 301.
79. Brandt, W. W., *ibid.*, p. 305.
80. Bierman, W. J., Gesser, H., *Analyt. Chem.*, 1960, **32**, 1525.
81. Keulemans, A. I. M., *Gas Chromatography 1960*, Editor Scott, R. P. W., Butterworth, London, 1960, p. 306.
82. Shipotofsky, S. H., Moser, H. C., *Analyt. Chem.* 1961, **33**, 521.
83. Abel, E. W., Nickless, G., Pollard, F. H., *Proc. Chem. Soc.*, 1960, 288.
84. Juvet, R. S., Wachi, F. M., *Analyt. Chem.*, 1960, **32**, 290.
85. Seely, G. R., Oliver, J. P., Ritter, D. M., *ibid.*, 1959, **31**, 1993.
86. Kaufman, J. J., Todd, J. E., Koski, W. S., *ibid.*, 1957, **29**, 1032.
87. Schomburg, G., *Gas Chromatography 1962*, Editor van Swaay, M., Butterworth, London, 1962, p. 292.
88. Ross, W. D., Scribner, W. G., Sievers, R. E., *Gas Chromatography 1970*, Editor, Stock, R., Institute of Petroleum, London, 1971, p. 369.
89. Sievers, R. E., *ibid.*, p. 378.
90. Pommier, C., *ibid.* p. 379.
91. Schwedt, G., Rüssel, H. A., *Chromatographia* 1972, **5**, 242.
92. Tourres, D. A., *ibid.*, p. 441.
93. Belcher, R., Jenkins, C. R., Stephen, W. I., Uden, P. C., *Talanta* 1970, **17**, 455.
94. Golay, M. J. E., *Gas Chromatography 1958*, Editor Desty, D. H., Butterworth, London, 1958, p. 36.
95. Ettre, L. S., Averill, W., *Analyt. Chem.* 1961, **33**, 680.
96. Desty, D. H., Haresnape, J. N., Whyman, B. H. F., *Analyt. Chem.*, 1960, **32**, 302.
97. Condon, R. D., *ibid.*, 1959, **31**, 1717.
98. Scott, R. P. W., Hazeldean, G. S. F., *Gas Chromatography 1960*, Editor Scott, R. P. W., Butterworth, London, 1960, p. 144.
99. Scott, R. P. W., *Nature*, 1959, **188**, 1753.
100. Desty, D. H., Goldup, A., *Gas Chromatography 1960*, Editor Scott, R. P. W., Butterworth, London, 1960, p. 162.

101. Turner, N. C., *Petrol. Refiner* 1943, **22**, 98.
102. Ray, N. H., *J. Appl. Chem.*, 1954, **4**, 21; *ibid.*, p. 82; *Analyst*, 1956, **80**, 863.
103. Janak, J., *Vapour Phase Chromatography 1956*, Editor Desty, D. H. Butterworth, London, 1956, p. 235.
104. McKenna, T. A., Idleman, J. A., *Analyt. Chem.*, 1960, **32**, 1299.
105. Greene, S. A., Pust, H., *ibid.*, 1957, **29**, 1055.
106. Szculczewski, D. H., Higuchi, T., *ibid.*, p. 1541.
107. Marvillet, L., Tranchant, J., *Gas Chromatography 1960*, Editor Scott, R. P. W., Butterworth, London, 1960, p. 321.
108. Scott, C. G., Phillips, C. S. G., *Gas Chromatography 1964*, Editor Goldup, A., Institute of Petroleum, London 1965, p. 266.
109. Phillips, C. S. G., *Gas Chromatography 1970*, Editor Stock, R., Institute of Petroleum, London 1971, p. 1.
110. Scott, C. G., *J. Inst. Petrol.* 1959, **45**, 118; *Gas Chromatography 1960*, Editor Scott, R. P. W., Butterworth, London, 1960, p. 137; *Gas Chromatography 1962*, Editor van Swaay, M. Butterworth, London, 1962, p. 36.
111. Halasz, I., *Gas Chromatography 1962*, Editor van Swaay, M., Butterworth, London, 1962, p. 133.
112. Scott, R. P. W., Lawrence, J. G., *J. Chrom. Sci* 1969, **7**, 65.
113. McCreery, R. L., Sawyer, D. T., *J. Chrom. Sci.* 1970, **8**, 122; *Analyt. Chem.* 1968, **40**, 106.
114. Kirkland, J. J., *Gas Chromatography 1964*, Editor Goldup, A., Institute of Petroleum, London, 1965, p. 285.
115. Genty, C., Schott, R., *Analyt. Chem.* 1970, **42**, 7.
116. West, D. L., Martson, A. L., *J. Amer. Chem. Soc.* 1964, **86**, 4731.
117. Petrova, R. S., Khrapova, E. V., Scherbakova, K. D., *Gas Chromatography 1962*, Editor van Swaay, M., Butterworth, London, 1962, p. 18.
118. Kirkland, J. J., *J. Chrom. Sci.*, 1970, **8**, 303.
119. Little, J. N., *ibid.* p. 647.
120. Hawton, I. J., Campbell, P., *ibid.* p. 675.
121. Eggertsen, F. T., Knight, H. S., Groennings, S., *Analyt. Chem.*, 1955, **27**, 170.
122. Petov, G. M., Scherbakova, K. D., *Gas Chromatography 1966*, Editor Littlewood, A. B., Institute of Petroleum, London 1967, p. 50.

123. Vidal-Madjar, C., Ganansia, J., Guiochon, G., *Gas Chromatography 1970*, Editor, Stock, R., Institute of Petroleum, London 1970, p. 20.
124. Goretti, G., Liberti, A., Nota, G., *Gas Chromatography 1968*, Editor Harbourn, C. L. A., Institute of Petroleum, London 1969, p. 22.
125. Brunnock, J. V., Luke, L. A., *Analyt. Chem.* 1968, **40**, 2158.
126. Hollis, O. L., *ibid.* 1966, **38**, 309.
127. Sakodynsky, K. I., Klinskaya, N. S., Panina, L. I., *ibid.* 1973, **45**, 1369.
128. Sie, S. T., Bleumer, J. P. A., Rijnders, G. W. A., *Gas Chromatography 1968*, Editor Harbourn, C. L. A., Institute of Petroleum, London 1969, p. 235.
129. Glueckauf, E., *Discuss. Faraday Soc.*, 1949, No. 7, 12.
130. Wilson, J. N., *J. Amer. Chem. Soc.*, 1940, **62**, 1583.
131. de Vault, D., *ibid.*, 1943, **65**, 532.
132. Glueckauf, E., *Trans. Faraday Soc.*, 1955, **51**, 34.
133. Keulemans, A. I. M., *Gas Chromatography*, Reinhold, New York, 1959.
134. Porter, P. E., Deal, C. H., Stross, F. H., *J. Amer. Chem. Soc.*, 1956, **78**, 2999.
135. Green, S. A., Pust, H., *J. Phys. Chem.*, 1958, **62**, 55.
136. Kwantes, A., Rijnders, G. W. A., *Gas Chromatography 1958*, Editor Desty, D. H., Butterworth, London, 1958, p. 128.
137. Anderson, J. R., Napier, K. H., *Austral. J. Chem.*, 1956, **9**, 541.
138. Hardy, C. J., *J. Chromatog.*, 1959, **2**, 490.
139. Evered, E., Pollard, F. H., *ibid.*, 1960, **4**, 451; Pollard, F. H., Hardy, C. J., *Analyt. Chim. Acta*, 1957, **16**, 135.
140. Phillips, C. S. G., *Gas Chromatography 1958*, Editor Desty, D. H., Butterworth, London, 1958, p. 340.
141. van Deemter, J. J., Zuiderweg, F. J., Klinkenberg, A., *Chem. Eng. Sci.*, 1956, **5**, 271.
142. Scott, R. P. W., *Gas Chromatography 1958*, Editor Desty, D. H. Butterworth, London, 1958, p. 189.
143. Bohemen, J., Purnell, J. H., *ibid.*, p. 6.
144. Golay, M. J. E., *Gas Chromatography 1960*, Editor Scott, R. P. W., Butterworth, London, 1960, p. 139.

Chapter Five

Thin-layer Chromatography

The idea of using a chromatographic adsorbent in the form of a thin layer fixed on an inert rigid support seems to have been suggested by Izmailov and Shraiber in 1938. Meinhard and Hall[1] in 1949 developed this notion of an 'open column', and in 1951 Kirchner, Miller, and Keller[2] reported the separation of terpenes on a 'chromatostrip', prepared by coating a small glass strip with an adsorbent mixed with starch or plaster of Paris, which acted as a binder. The strips were handled in the same way that paper is handled in paper chromatography, and indeed the original object of the thin-layer technique was to apply the methods of paper partition chromatography to an adsorption system.

Several years passed before the method became widely used, probably because at that time the development of paper and gas chromatography was proceeding rapidly. In the late 1950's, however, Stahl[3, 4] devised convenient methods of preparing plates, and showed that thin-layer chromatography could be applied to a wide variety of separations. He introduced a measure of standardisation, and since the publication of his work and the appearance of commercial apparatus based on his designs many more workers have taken up the method. As usually happens, once the initial stimulus has been given, many variations of the original procedures have been proposed and the use of thin-layer chromatography has increased so rapidly that in many fields it has superseded paper chromatography, chiefly because it is usually quicker, and gives better separations. Its use has also enabled the simple techniques of paper chromatography to be used for separations which do not work on paper.

279

Outline of the method

It is first necessary to prepare the chromatographic plates; that is, to spread the adsorbent in a thin film of even thickness on a firm inert support. Glass plates are commonly used, but alternatives are mentioned later (p. 311). The finely powdered solid adsorbent is usually made into a slurry with water (less commonly with a volatile organic liquid), and spread on the plate by means of commercial spreading apparatus or some simple 'home-made' spreader or even by the exercise of manual dexterity alone. It is also possible to make the layer by a spraying or dipping technique. The spread plate is dried and 'activated' by heating at about 100°C for a pre-determined period.

A solution of the sample in a volatile solvent is applied with a pipette or syringe (or other method as described for paper chromatography on p. 143). When the spot has dried the plate is placed vertically in a suitable tank with its lower edge immersed in the selected mobile phase, and an ascending chromatographic separation is thus obtained. If a more complicated apparatus is used, descending or horizontal (circular) chromatography can be carried out, but these are less common. At the end of the run the solvent is allowed to evaporate from the plate, and the separated spots are located and identified by physical or chemical methods, as used for paper chromatograms. It will be seen that, apart from the preparation of the plates, the practical technique required is very similar to that of paper chromatography, and in the more detailed descriptions of the various steps which follow, frequent reference will be made to the corresponding parts of Chapter 3.

Comparison of thin-layer with other forms of chromatography

To assess the relative value of the thin-layer method, it is necessary to compare it first with adsorption column chromatography, because the same *physical* system is being used in both cases, and second with paper partition chromatography, because the same *experimental* techniques are used in both.

Conventional column chromatography is a fairly slow process which requires relatively large amounts of adsorbent and sample. Recent developments with high efficiency systems (described on p. 85) have certainly overcome the problems of speed, scale, and identification, but they carry the penalty of greatly increased cost and complexity of the apparatus. In thin-layer chromatography only small amounts of adsorbent and minute samples are needed, and the apparatus is simple and fairly cheap. The separated spots are located on the plate in the manner of paper chromatography, so that normally no collection of fractions is necessary. There is, however, no difficulty about preparative separations (Preparative Layer Chromatography), which are achieved by increasing the thickness of the layer and using a higher loading of sample. After separation, it is easy to recover an individual substance by scraping off and collecting the part of the layer on which it is adsorbed. The substance can then be extracted with a suitable solvent.

In comparison with paper chromatography, the thin-layer method has the main advantage of greater speed, and in most cases, better resolution. The average time for a 10 cm run in thin-layer chromatography on silica gel is 20–30 minutes (depending on the nature of the mobile phase), whereas the same separation on a fast paper might take two hours. Rough qualitative separations on small plates may take as little as five minutes. The better resolution arises from the fact that the adsorbent in thin-layer chromatography has a higher capacity than the paper in paper chromatography. The separated spots therefore retain fairly closely the shape and size of the original applied spot, without the spreading associated with partition chromatography on paper. This advantage is largely lost when a partition system is used on a thin layer. A further, and very important, advantage of the adsorption system is that it can be used to separate hydrophobic substances, such as lipids and hydrocarbons, which are difficult to deal with on paper, even with a reversed phase system. Thin-layer separations have been applied, however, in most fields of organic, and some of inorganic, chemistry.

Location of separated substances on thin layers is done in the same way as it is on paper, but more reactive reagents, such as concentrated sulphuric acid, can be applied on thin layers, provided that the thin-layer material is an inert substance such as silica gel or alumina.

Two minor disadvantages of thin-layer as compared with paper chromatography are the greater difficulty of recording and preserving thin-layer chromatograms, and the fact that it is not quite so easy to achieve reproducibility of R_f values. Both disadvantages can be minimised, and they are between them not enough to reduce the almost overwhelming superiority which thin-layer chromatography is tending to achieve over all other forms except gas chromatography.

Adsorbents

The general properties of adsorbents for thin-layer chromatography should be similar to those described on p. 27 for adsorbents used in columns, and the same arguments about 'activity' apply. Two important properties of the adsorbent are its particle size and its homogeneity, because adhesion to the support largely depends on them. A particle size of $1-25$ μm is usually recommended. A coarse-grained material will not produce a satisfactory thin layer, and one of the reasons for the greatly enhanced resolution of thin-layer chromatography is this use of a very fine-grained adsorbent. Whereas in a column a very fine material will give an unacceptably slow flowrate, on a thin layer the fine grain gives a faster and more even solvent flow. Some examples of adsorbents which have been used for some representative separations by thin-layer chromatography are given in Table 5.1.

Table 5.1 *Adsorbents for thin-layer chromatography*

Solid	Used to separate
Silica gel	Amino-acids,[5] alkaloids,[6] sugars,[7] fatty acids,[8,9] lipids,[10] essential oils,[11], inorganic anions and cations,[12] steroids,[13,14] terpenoids[15]
Alumina	Alkaloids,[6] food dyes,[16] phenols.[17] steroids,[14,18] vitamins,[19] carotenes,[20] amino-acids[59]
Kieselguhr	Sugars,[7] oligosaccharides,[21] dibasic acids,[22] fatty acids,[23] triglycerides,[24] amino-acids,[25] steroids[26]
Celite	Steroids,[27] inorganic cations[1]
Cellulose powder	Amino-acids,[28,29] food dyes,[30] alkaloids,[31] nucleotides[32]
Ion-exchange cellulose	Nucleotides,[33] halide ions[34]
Starch	Amino-acids[35]
Polyamide powder	Anthocyanins,[36] aromatic acids,[36] antioxidants,[37] flavonoids,[38] proteins[39]
Sephadex	Amino-acids,[40] proteins[41,42]

Silica gel. The most commonly used adsorbent is silica gel, which can be prepared by the methods described in Chapter 2. It is, however, laborious to prepare the gel, and commercial materials are in most cases entirely satisfactory.

In most of the work reported on silica gel plates a binder is used, to give more mechanical strength to the layer, and to improve adhesion to the glass support. The binder most often used is calcium sulphate (plaster of Paris), which is intimately mixed with the silica gel in the proportion of about 10% by weight. It is important that the binder and the silica shall be thoroughly mixed so that the layer is absolutely homogeneous, and it is probably better to use a commercial silica gel ready mixed with binder than to mix them oneself. Silica

Gel G, although a trade name, seems now to be universally used to denote silica gel with a calcium sulphate binder.

Although the presence of calcium ions does not affect most separations, and layers with binder are much easier to handle, silica gels are now obtainable which adhere sufficiently without a binder, and are therefore preferred by some workers. Another binder which has found a limited use is starch, but it places restrictions on the use of corrosive locating reagents.

Alumina. Although alumina is historically the most important adsorbent for chromatography, it has not been extensively used in thin-layer chromatography. The properties required are the same as for column chromatography, and the factors controlling the type of layer obtained are similar to those for silica gel. Alumina may be used with a binder, but some manufacturers have given special attention to the production of alumina in a suitable form without a binder, and the use of an unbound layer is more common with alumina than with silica gel. Alumina may be obtained in acidic, neutral, or basic forms, and it is more useful than silica when a high or low acidity adsorbent is wanted.

Cellulose powders. It may at first seem unnecessary to go to the trouble of preparing cellulose powder plates when paper could be used more conveniently, but there are important advantages in the thin-layer method. In paper (p. 124) the cellulose is fibrous, and the fibres, however closely they are matted together, inevitably form a network with large gaps. The solvents flow along the surface of the fibres, and the gaps become filled with liquid, with the result that excessive diffusion of solutes takes place, and the separated zones tend to be larger than the original spot. If the fibres are too tightly compressed the flowrate becomes unacceptably slow.

Layers of cellulose powder, on the other hand, are aggregations of very small particles, all of much the same size. The interstices are therefore much smaller and more regular, and the adsorbent surfaces are more evenly distributed. In consequence there is a much more even flow of mobile phase, with less diffusion of the dissolved substances. The flow is also much faster.

Modified cellulose powders (p. 158) can be used to obtain ion-exchange separations on thin layers, with similar advantages over column or paper sheet methods. Both normal and modified cellulose powders can be used without a binder. *Molecular sieve layers.* Thin layers on glass plates can be made from Sephadex gel chromatography media (p. 73), with a superfine grade having a particle size $10-40\,\mu$m. Chromatography with these layers is rather slower and more troublesome than other forms of thin-layer chromatography, but it may be faster and more convenient than gel chromatography in columns.

Other adsorbents. Most of the adsorbents mentioned in Table 2.1 have been tried at various times for making thin layers. Others are kieselguhr, Celite, calcium sulphate and polyamide (polycaprolactam) powders. Limited uses of most of these have been reported with and without a binder (calcium sulphate plates are sufficiently cohesive to be washed like paper).

All the adsorbents so far mentioned can be used for adsorption chromatography, but some of them, such as silica gel, Celite, kieselguhr, or cellulose, can be used for thin-layer partition chromatography (as they can be used for partition chromatography in columns, p. 35), if an appropriate mobile phase is selected. If the plate is dried so that it retains very little adsorbed water, and the mobile phase is a non-polar mixture, separation will be by adsorption chromatography. If, however, the plate retains much water, or the solvent mixture contains a highly polar constituent, separation will be largely by partition chromatography. Cellulose powder is usually used to provide a partition system (which limits the advantages of the material over paper sheets), and modified cellulose gives an ion-exchange separation, with an aqueous buffer as the mobile phase.

Preparation of plates

The adsorbent is spread on a suitable firm support, which may be quite rigid, or flexible. Glass plates were the original support and are still the most usual; flexible plates are described later. The size used depends on the type of separation to be carried out and on the type of chromatographic tank and spreading apparatus available. Microscope slides are often very convenient for rapid qualitative separations, and old photographic (¼ plate) plates, or spectrograph plates can be used. Most of the commercial apparatus is designed for plates of 20 x 5 or 20 x 20 cm, and those are now regarded as 'standard'. It is important that the surface of the plate shall be flat, and without irregularities or blemishes.

Glass plates are cleaned thoroughly before use, washed with water and a detergent, drained, and dried. A final wash with acetone may be included, but it is not essential. It is important not to touch the surface of the cleaned plates with the fingers.

The first step is to make the adsorbent into a slurry with water, usually in the proportion x g of adsorbent and $2x$ cm^3 of water. The slurry is thoroughly stirred, and spread on the plate by one of the methods described below. If a binder is used, the time available from mixing the slurry to completion of spreading is about four minutes (after which setting will have begun). The properties of the adsorbent can be modified by using buffer solutions instead of water, to give a layer of desired acidity, or to modify the water-retaining properties of the layer. Similarly, complexing agents or fluorescent indicators can be incorporated in the layer by mixing the slurry with solutions of the appropriate substances. An important example is the use of silver nitrate in the separation of lipids and related materials.

The film thickness is a most important factor in thin-layer chromatography. The 'standard' thickness is 250 μm, and there is little to be gained by departing very much from that in analytical separations (thinner layers may give rise to erratic R_f values – p. 304). Thicker layers (0.5 to 2.0 mm) are used for preparative separations, with a loading of up to 250 mg on a 20 x 20 cm plate. One difficulty with thick layers is their tendency to crack on drying.

There are in principle four ways of applying the thin layer to its support: spreading, pouring, spraying, and dipping.

Spreading. It is possible to spread the adsorbent in a number of ways, but the main objective is to produce an absolutely even layer with no lumps or gaps, which adheres evenly and securely to the support.

The original method of spreading was by means of a spatula, and a number of variations of that manual technique are described below. It is not essential to use a commercial spreader, but it is very much easier, and unless cost is an important consideration 'do-it-yourself' methods are not necessarily better.

Commercial spreaders are in general of two types, which might be called the 'moving spreader' and 'moving plate' types. The difference is shown diagrammatically in Figs. 5.1 and 5.3. Both types are made in varying degrees of sophistication (and cost), and since each maker supplies detailed instructions for the use of his equipment, only the general principles are mentioned here.

In the 'moving spreader' type, the glass plates (usually five 20 x 20 cm, or an equivalent number of smaller size) are held in a flat frame, and a rectangular hopper containing the slurry is passed over them. The hopper has no bottom (or the bottom can be opened when required), and its trailing face has an accurately machined lower edge to give an even layer of the required thickness. In some appliances the thickness is adjustable. The lower edge of the leading face of the hopper rests on the glass plates.

It will be seen that an arrangement such as that shown in Fig. 5.1 requires that all the plates shall have the same thickness, so that the leading edge of the hopper does not catch on the edges of the plates. It is not difficult to select suitable plates, but to make selection unnecessary some makers have produced spreading 'beds' in which the long sides have an overhanging lip (Fig. 5.2). The plates are pushed

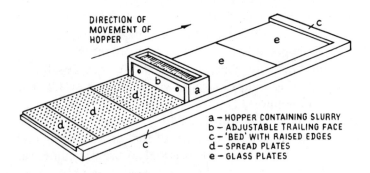

DIRECTION OF MOVEMENT OF HOPPER

a – HOPPER CONTAINING SLURRY
b – ADJUSTABLE TRAILING FACE
c – 'BED' WITH RAISED EDGES
d – SPREAD PLATES
e – GLASS PLATES

Fig. 5.1. 'Moving spreader' apparatus

from below against the lip, by means of spring strips, or in one case, by an inflatable air-bag, so that the upper surfaces of the plates are at the same level, regardless of their thickness. A foam-rubber base on which the plates are laid has also been suggested (van Damm and Maas[43]) as another way of getting an even spreading surface. With a little practice the 'moving spreader' method is a very satisfactory way of making thin-layer plates.

In the 'moving plate' apparatus (Fig. 5.3) the hopper is fixed to the middle of the bed. It has one fixed face, and one which is adjustable to give layers of different thickness. A plate is put under the hopper, slurry is poured in, and a second plate is used to push the first through under the hopper. Successive plates are then pushed through until the hopper is empty. Metal strips are fixed on the sides of the

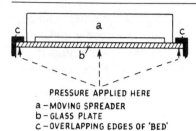

a – MOVING SPREADER
b – GLASS PLATE
c – OVERLAPPING EDGES OF 'BED'

*Fig. 5.2. Modified bed for moving
spreader apparatus*

bed on either side of the hopper to guide the plates through. In this apparatus all the plates must have the same thickness, and they must also all have as nearly as possible the same width, as they will jam between the guides if they are too wide, and if they are too narrow they tend to twist as they go through, and jam in that case also.

A further disadvantage of the 'moving plate' apparatus is that the plates must be lifted from the bed as soon as they have been spread, to allow the remainder of the set to pass under the hopper. In the 'moving spreader' apparatus, the whole batch of plates may be left undisturbed on the bed until they have dried a little. They can then be moved with less risk of damage.

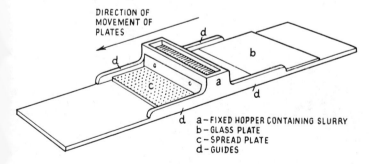

a – FIXED HOPPER CONTAINING SLURRY
b – GLASS PLATE
c – SPREAD PLATE
d – GUIDES

Fig. 5.3. 'Moving plate' apparatus

Numerous other methods of spreading have been suggested. A very simple method is to use as the spreader a glass rod which has two strips of cellulose tape wrapped round it, the same distance apart as the width of the plate (Fig. 5.4).

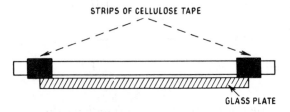

Fig. 5.4. Glass rod for spreading

Alternatively[44], strips of tape can be put along the edges of the plate, and the slurry can be spread with a glass rod or a ruler resting on the tape (Fig. 5.5). The tape should be removed after spreading; ordinary transparent cellulose tape is best avoided in this case, because it is not always easy to remove it from glass, it is not waterproof, and it may leave traces of adhesive behind.

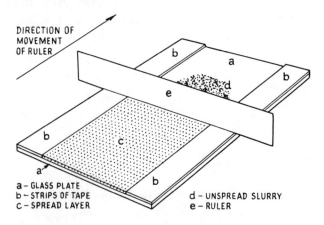

Fig. 5.5. Use of strips of tape to control layer thickness

A variation of the last method is illustrated in Fig. 5.6. Three plates of the same thickness are put side by side on a wood base which has one slightly raised edge to locate them. The centre plate is the one to be spread, and the others act as rests for the spreader, which is a ruler or a glass rod. The thickness of the layer is obtained by raising the side plates on strips of paper (for example, strips of Whatman 3MM chromatography paper).

In all these simple methods a suitable amount of the slurry should be placed at one end of the plate, and it should be spread in one firm movement of the spreader. If a binder is being used, any attempt to smooth out blemishes by further passes of the spreader will only make matters worse. If no binder is used, several passes of the spreader may be made, but it is still better to make only one, and a satisfactory technique will soon be gained with a little practice.

Pouring. Many workers prefer not to use mechanical spreading methods at all. If the adsorbent is very finely divided and of homogeneous particle size, and if no binder is used, a slurry can be poured on a plate and allowed to flow over it so

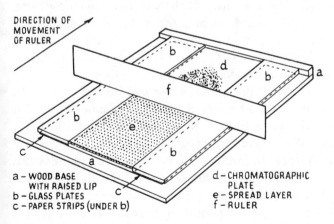

DIRECTION OF MOVEMENT OF RULER

a – WOOD BASE WITH RAISED LIP
b – GLASS PLATES
c – PAPER STRIPS (UNDER b)
d – CHROMATOGRAPHIC PLATE
e – SPREAD LAYER
f – RULER

Fig. 5.6. A simple method of spreading

that it is evenly covered. Some manual dexterity is required to do this properly. Preparation of plates by pouring is particularly easy with certain types of alumina, but water alone is not usually suitable for making the slurry; a volatile liquid such as ethanol (or an ethanol-water mixture) or ethyl acetate is preferable. The appropriate amounts of liquid and solid adsorbent needed to cover a plate have to be found by trial and error, and exactly those quantities should be used to ensure that the thickness of the layer is reproducible. Good even plates can be made by this method, but the thickness of the layer is not known.

Spraying. Descriptions or details of methods of making plates by spraying with a slurry of adsorbent have been published (for example, by Bekersky[45], and by Morita and Harata[46]). The methods do not seem to have any particular advantage over spreading, and are open to the objections usually associated with spraying, among which are the difficulty of getting even coverage, and the fact that there is no easy way of ensuring reproducible thickness.

Dipping. Small plates, such as microscope slides, can be spread by dipping in a slurry of the adsorbent in chloroform, or other volatile liquid. Again, the exact thickness of the layer is not known, and the evenness of the layer may not be very good, but this is a most convenient method for making a number of plates for rapid qualitative separations.

After spreading, the plate is allowed to dry for 5–10 minutes, and, if it has been made with an aqueous slurry, it is further dried and 'activated' by heating at about 100°C for 30 minutes. Plates made with volatile organic liquids may not need this further drying. It is important to standardise this part of the preparation of plates, because the activity of the adsorbent may depend rather critically on it (see p. 303).

Plates may be kept for short periods in a desiccator, but long storage is not recommended. When inspected for imperfections before use, they should appear uniform in density when viewed by transmitted or reflected light, and there should be no visible large particles. Gentle stroking with the finger should not remove the layer, and there should be no loose particles on the surface.

The mobile phase

Choice of the mobile phase depends on much the same factors that govern the choice for adsorption chromatography in columns (p. 26). It is preferable to use an organic solvent mixture of as low polarity as is consistent with a good separation. One reason for so doing is to minimise adsorption of any components of the solvent mixture; if highly polar components (particularly water) are included in the mixture, sufficient may be taken up to convert the system into a partition one. It is usually better to avoid this, although in certain cases a better separation may thus be obtained. An additional reason for avoiding the use of water is that it may loosen the adhesion of a layer on a glass plate, and thus give rise to mechanical imperfections in the separation.

In selecting a solvent, the 'eluotropic series' referred to on p. 27 can be consulted; it has been found that they are applicable to thin-layer in much the same way as to column separations. Suitable mixing gives mobile phases of intermediate eluting power, but it is best to avoid mixtures of more than two components as much as possible, chiefly because more complex mixtures readily undergo phase changes with changes in temperature. When mixtures are used, greater care is necessary over equilibration (p. 305). The purity of the solvents is of much greater importance in thin-layer than in most other forms of chromatography, because of the small amounts of material involved.

Application of samples

Samples can be applied to thin layers by any of the methods of paper chromatography (p. 143), but considerably more skill and care are needed. The pipette, loop, or syringe can be allowed to rest lightly on a paper surface while the solution soaks in, but if the same thing is done on a thin layer there is every likelihood that a hole will be made right through the adsorbent layer. A hole is most undesirable, since its presence is a major obstruction to even solvent flow. The irregularity of solvent flow round a hole causes the moving spot to be completely distorted, and may prevent the separation of substances of close R_f value.

It is possible, with practice and a steady hand, to apply samples with a platinum loop, but a capillary tube or a micro-pipette is better. The emerging drop should just touch the surface, and the tip of the tube should remain just above it. A Hamilton glass syringe (or an Agla syringe) is the best instrument to use for spotting plates. It is not necessary for the tip of the syringe to touch the surface, although it should be held as close to it as possible.

The solvent in which the sample is dissolved for spotting should be as volatile as possible, and also have as low a polarity as possible. If the spotting solvent is strongly adsorbed by the layer, marked irregularities may be observed as the mobile phase passes the position of the spots, and the separated spots may be seriously distorted.

Individual spotting positions cannot be marked with a pencil, as is done on paper, and therefore a template, or spotting guide, should be used. All that is needed is a Perspex sheet with strips cemented along two opposite edges slightly thicker than the glass plates (Fig. 5.7), which is placed over the chromatographic plate with the edge along the start line. Marks on the template 1 cm apart indicate where the spots should be placed. Templates can be bought for 20 x 20 cm

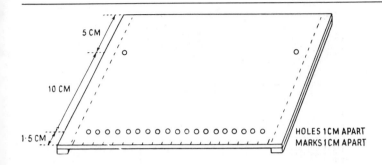

Fig. 5.7. Spotting template

plates which have, in addition to guide marks for spotting, lines for marking out the finishing line, and for reading off R_f values. Apart from being an aid to accurate spotting, the template acts as a rest for the hand, and shields the surface of the plate from damage.

Spots should not be nearer than 1 cm centre to centre, they should be 2–5 mm in diameter and should not be nearer to the edge of the plate than 1.5 cm on a 20 x 20 cm plate. A volume of 1–5 mm³ should be applied, and if a larger volume of the sample solution than that is needed to give the required loading, it should be applied in portions (as on paper, p. 144). Several workers (for example Morgan[47], and Curtis[48]) have designed simple devices to facilitate loading in portions and also to enable a number of spots to be applied to the plate at the same time. The loading should not be more than about 15 μg per spot on a layer 250 μm thick, 10 μg being the optimum amount for most substances. In preparative work the sample is often applied as a streak along the start line; up to about 4 mg may thus be loaded on a 20 x 20 cm plate 250 μm thick. Mechanical devices can be obtained for use in conjunction with a syringe to give rapid and even streaking.

The start line can be marked at the edge of the plate with a sharp pencil, as shown in Fig.5.8, and the finishing line can

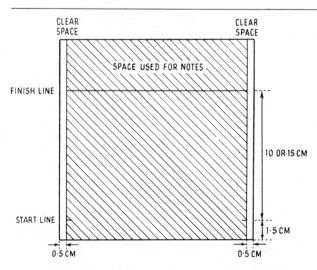

Fig. 5.8. Plate marked out for one-way ascending chromatography

be drawn right across the plate, with the edge of the template as a guide. The pencil removes a fine line of adsorbent down to the glass, and the solvent flow is forced to stop when the front reaches the line, a useful feature which enables one to standardise conditions very easily. The plate should not, however, be left standing in the tank for a long time after the front has reached the stop line, as diffusion and evaporation may cause spreading of the separated spots (see p. 305). Details of the run, solvent, solutes, and so on, can be written with a pencil in the unused spaces. The edges of the plate should be rubbed clear with a finger before spotting to a width of about 0.5 cm, to give a sharper edge to the adsorbent layer. The edge to be immersed in the mobile phase should not be allowed to become damaged. For a two-way separation two adjacent edges must be left untouched, and the plate marked out as shown in Fig. 5.9. If it is desired to ensure that the spots do not diffuse or move sideways, and thus interfere with each other, lines may be

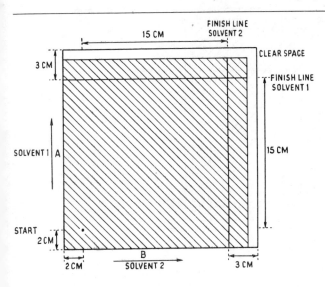

Fig. 5.9. 20 × 20 cm plate marked out for two-way ascending chromatography

drawn with a pencil as shown in Fig. 5.10, to give what is in effect a set of separate strip chromatograms.

Running

When the chromatographic plate has been prepared, and the samples have been applied to it, it is placed in a suitable tank with the lower edge immersed in the selected mobile phase to a depth of 0.5–1.0 cm. Two plates can usually be put in one tank (Fig. 5.11). The lid of the tank must fit securely. There is often no need to allow any equilibration time (but see p. 305), but to ensure homogeneity of the atmosphere in the tank the walls may be lined with sheets of filter paper soaked in the mobile phase.

The smallest possible tank should be used, so that the enclosed atmosphere has the smallest possible volume. It is not essential to use an orthodox tank at all. A 'sandwich' can

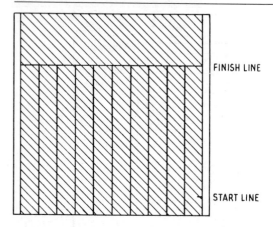

Fig. 5.10. Plate marked to give separate strips

be made, consisting of a second glass plate, the same size as the chromatographic plate, clamped firmly over the adsorbent surface, with a glass, metal, rubber, or plastic spacer at the edges to protect the layer from damage (Fig. 5.12). The surface of the solvent in the trough should not be exposed to the external atmosphere, or there will be loss of volatile components by evaporation. There is a commercial version of this 'sandwich' apparatus.

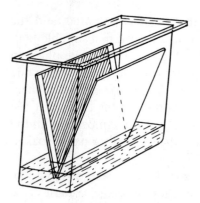

Fig. 5.11. Small tank for two 20 × 20 cm plates

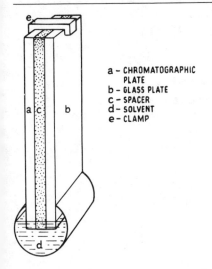

a – CHROMATOGRAPHIC
    PLATE
b – GLASS PLATE
c – SPACER
d – SOLVENT
e – CLAMP

Fig. 5.12. 'Sandwich' type apparatus

For plates of non-standard size any suitable closed vessel can be used; for example, a beaker (without pouring lip) covered with a clock glass, or a gas jar covered with a glass disc.

Sometimes the adsorbent tends to flake off the plate below the solvent surface. This usually happens when a highly polar solvent is being used, and traces of grease on the plates will assist it. It is of no significance as long as the lower edge of adhering adsorbent remains in contact with the liquid, but it must be remembered that the level of the solvent surface will be tending to sink at the same time.

If descending or horizontal development is desired, a more complicated apparatus is necessary. An arrangement for descending chromatography is shown in Fig. 5.13, and for horizontal development in Fig. 5.14. In the method of Fig. 5.14 (*a*) the development is similar to that obtained in the ordinary ascending or descending method, but in the method of Fig. 5.14 (*b*) radial development occurs. In the last

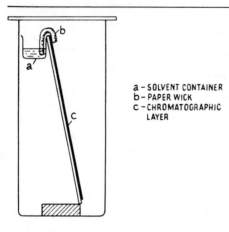

a – SOLVENT CONTAINER
b – PAPER WICK
c – CHROMATOGRAPHIC
    LAYER

Fig. 5.13. Descending thin-layer chromatography

example a hole is cut in the plate to accept a paper or cotton wick. In a variation proposed by Litt and Johl[49] the plate is placed with the layer facing downward with its centre in contact with a porous wick pushed upward by a plastic spring standing in the solvent. The procedure is similar to that for radial paper chromatography (p. 118 and Fig. 3.13).

The object of radial development is to make use of the 'concentration' effect of spreading the zones out into a thin line, rather than a spot, and thus separate substances of close R_f values. The object of descending development is usually to get an apparent lengthening of the run by allowing the solvent to run off the bottom of the plate. Substances of low R_f are thus moved further along the plate. If neither of these two effects is needed, there is little point in departing from the usual ascending development. On the other hand, if the run is to be carried out below room temperature, the type of apparatus which lends itself most readily to the introduction of cooling devices is that shown in Fig. 5.14 (*a*).

After the run the plate is taken from the tank and allowed to dry, without heating. A gentle current of air may be

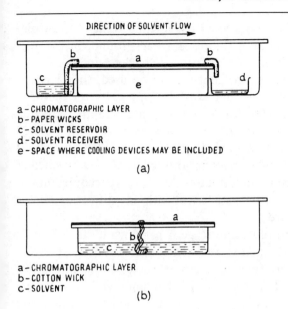

a – CHROMATOGRAPHIC LAYER
b – PAPER WICKS
c – SOLVENT RESERVOIR
d – SOLVENT RECEIVER
e – SPACE WHERE COOLING DEVICES MAY BE INCLUDED

(a)

a – CHROMATOGRAPHIC LAYER
b – COTTON WICK
c – SOLVENT

(b)

Fig. 5.14. Horizontal thin-layer chromatography
(a) One-way development
(b) Radial development

passed horizontally over the surface. As in paper chromatography, in two-way separations the first solvent must be removed completely before the run in the second.

Location

In general, the methods of location of colourless substances on thin layers are similar to those used in paper chromatography. Physical methods include ultra-violet fluorescence and radioactive counting. As with paper, it should be a routine to examine the plate under ultra-violet light, when the spots may be revealed by their fluorescence. Alternatively, the whole plate may be made fluorescent at the beginning, either by inclusion of a suitable dye in the solid adsorbent, or by using a solution of the dye to make the slurry which is spread on the plate. The spots are then located under ultra-violet light as dark spots on a fluorescent background, or by their fluorescing a different colour from the background.

If chemical locating reagents are to be used, they have to be applied by spraying, and not by dipping, by the methods described on p. 150 for paper chromatography. Organic substances may be located by spraying with concentrated sulphuric acid, and then heating to about 200°C for about ten minutes. The substances are thus charred, and revealed as dark spots. This is an effective locating procedure, but it gives, of course, little assistance with identification, being entirely unselective.

Another general locating reagent for organic substances is iodine. The plate is exposed to iodine vapour, by being placed for a short time in a closed vessel containing a few iodine crystals. The background becomes violet, with unsaturated substances showing up as colourless spots, and many saturated substances as brown spots. All the colours fade rather quickly on exposure to the atmosphere. This is not a very good locating reagent, but, like the fluorescence method, it has the advantage that most of the substances are not chemically changed by it, and, their positions having been marked, they are readily eluted from the plate. Apart from these general methods, most of the locating reagents used on paper for specific compounds or groups of compounds are applicable to thin layers in the same way.

Identification and R_f values

Identification of the separated substances on thin layers, if not done by elution and individual tests on the eluates, is best achieved by means of chemical locating reagents and colour reactions, although the relative positions of two or more substances on the same chromatogram can often be used with some confidence. As in paper chromatography, it is best to run reference substances or mixtures on the same plate as the unknown.

R_f values, defined on p. 153, are rather less valuable in thin-layer chromatography than they are on paper, because in the former case there are additional variable factors which have an influence on chromatographic behaviour. The factors which are operative on thin layers are:

The nature of the adsorbent. Different adsorbents will clearly give widely different R_f values with the same mobile phase and solutes, but reproducibility will only be obtained with a given adsorbent if the particle size is constant, and if the binder (if any) is homogeneously mixed.

The mobile phase. The purity of the solvents used in the mobile phase is very important in thin-layer chromatography, and when mixed solvents are used the proportions should be fairly rigorously controlled. There is a tendency to use rather volatile liquids on thin layers, which makes this control more difficult but at the same time more important.

The activity of the adsorbent. In an adsorption system the activity of the adsorbent (defined on p. 28) clearly plays a major part in determining R_f values, not only because of the adsorptive power for the solutes, but also because of the effects on a mixed mobile phase. There is bound to be some differential adsorption of solvent components, giving in effect a gradient elution, or even frontal analysis of the solvent mixture, and the eluting power of a particular solvent mixture will be different on adsorbents of different activity. In a partition system the affinity of the adsorbent for water or polar solvent constituents is the important factor.

The objective is to find a way of preparing plates of known and reproducible activity. It is possible to start with a material of known activity, particularly in the case of alumina, whose activity on the Brockmann scale can be estimated, but that does not give the activity of the final plate, which depends largely on the amount of water retained after spreading and drying.

Rigorous control of storage conditions of the dry adsorbent, and of the whole process of preparation, pretreatment, and storage of plates, helps to reduce the effects of variations in the activity factor, but in spite of this it has been recommended by Stahl that a test mixture of three coloured substances should be run on every chromatogram, as a check on the activity of the adsorbent. The mixture is of the three dyes indophenol blue, Sudan red G, and p-dimethylaminoazobenzene (Butter yellow). It has even been suggested that the last substance should be used as a reference standard, and that R_b values, relative to its position on the chromatogram, should be quoted instead of R_f values.

The acidity of the adsorbent has a considerable effect on the relative R_f values of acidic and basic substances. Variations in acidity are most likely to be encountered when alumina is the adsorbent.

The thickness of the layer. It has been found that in the region 200–300 μm the thickness of the layer is not critical, and the 'standard' thickness of 250 μm is based on that observation. Below 200 μm R_f values may vary considerably, and for an individual substance they may be higher or lower than normal. The effect of layer thickness has been examined by Pataki and Kelemen[50]. It should be noted that the exact thickness of the layer as actually used is not known. The accurately measured figure is the setting of the spreading apparatus, and the wet layer presumably has that thickness. On drying and activation some shrinkage probably takes place, which is another reason, besides the activity consideration, for standardisation of procedure. There could be some slight swelling of the dry layer when it comes into contact with the mobile phase, but possibly not to the original wet thickness.

Although in practice the thickness may not have a marked effect, it is still necessary that the thickness shall be even, because thin or thick patches will cause uneven solvent flow in small areas of the plate.

The temperature. Separations should be carried out at a reasonably constant temperature, mainly to avoid changes in solvent composition caused by evaporation or phase changes. Pure adsorption separations with a one-component mobile phase are not much affected by changes in temperature of up to $\pm 10°$C.

Equilibration. It has been shown that equilibration is more important in thin-layer than in paper chromatography, and therefore it is important to keep the atmosphere in the tank saturated with solvent vapour. If the lid fits properly, and the vessel is as small as possible, saturation can be assisted by lining the walls with filter paper soaked in solvent. With these precautions, however, it is not usually necessary to allow any 'equilibration time' before the start of the chromatographic run. The size of the tank and the actual volume of a mixed solvent used have an effect on R_f values, because between them these two factors control the composition of the vapour in the tank. This in turn controls the rates of evaporation of solvents from the plate during the run. Standardisation of the geometrical parameters of the plate is desirable; for instance, length of run, distance of the start line from the solvent surface (or position of solvent feed), and positions of initial spots, should be kept constant. An indication that the atmosphere in the tank is not saturated with vapour, when a mixed solvent is being used, is the development of a concave solvent front, the mobile phase advancing faster at the edges than in the middle. This condition should be avoided.

Loading. The best results are obtained with a loading of about 10 μg per spot on a 250 μm plate. Loadings greatly in excess of that tend to cause spreading of the spots, with possible tailing and other non-equilibrium effects, and thus lead to erratic R_f values. The relative amounts of different substances in a mixture are also important, as the substances can interfere with each other and the R_f value of a substance by itself will not necessarily be the same as its R_f value in a mixture.

The experience of workers in different fields is that in some cases reproducibility of R_f values is difficult to achieve, whereas in others it is relatively easy. This seems to suggest that in thin-layer, as in paper, chromatography, there is a large 'personal' factor which has a bearing on the results obtained, and it reinforces the importance of standardisation of procedure, as recommended by Stahl. The subject of R_f values in thin-layer chromatography has been reviewed by Shellard[51] and by Dallas[52].

Quantitative determination

The methods used for quantitative evaluation of separated substances are similar to those used on paper. Radioactive substances are the easiest to deal with, as they can be scanned on the plate with a suitable counter. For other substances, an estimate can be made from the area of the spot, which has been shown to bear a simple relationship to the amount of material present.

If there is a suitable quantitative colour reaction, the spot density can be measured with a photodensitometer which can measure reflected light. The opacity of the layers is too great for transmitted light to be of any value in measurements.

A third method is to remove the part of the layer containing the substance, and to elute it separately. Removal can be effected with a spatula, which is used to scrape off the relevant part of the layer, or with a suction device, which draws the adsorbent into a small receiver (Fig. 5.15). Mechanical scrapers which remove selected parts of the layer are available commercially.

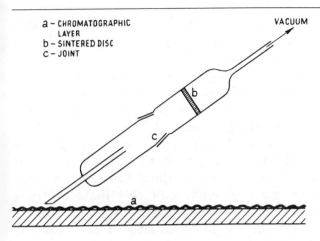

a – CHROMATOGRAPHIC LAYER
b – SINTERED DISC
c – JOINT
VACUUM

Fig. 5.15. Apparatus for removal of part of a thin-layer chromatogram

Recording of thin-layer chromatograms

One of the advantages of paper chromatography is that the paper sheet itself can be annotated and filed as a permanent record (although, since the colours usually fade or change rather quickly, this is of limited value). It is hardly possible to preserve thin layers on glass plates, and therefore sketching or photography are the usual ways of recording thin-layer chromatograms.

It is, however, possible to treat the layer with a binding agent and then to peel it off the support. It then looks rather like a paper sheet. The binding agent is polyvinyl propionate, which is sprayed on the chromatogram and allowed to dry. The plate is then soaked in water until the layer can be detached.

Variations of thin-layer chromatography

So far in this chapter a 'standard' system of thin-layer chromatography has been described, with a firm homogeneous layer spread evenly on a rigid support. Many variations have been proposed, mostly to extend the convenience of the method, or its range of applications, and an outline of a few now follows.

Loose-layer chromatography. Many workers have found advantages in spreading a layer of dry adsorbent on a plate by one of the simple methods described on p. 290, possibly because this method removes some of the uncertainties about the activity of the layer. The plate is spotted in the usual way, but development must be in a horizontal (or nearly horizontal) position. Locating reagents, if any are used, must be applied by spraying a fine mist over the plate (lying horizontally) so that it settles without disturbing the layer.

Gradient plates. The use of plates made in the usual way (with or without a binder) but whose adsorptive properties change across the layer was suggested by Stahl[53], who designed a gradient spreader (of the moving spreader type) for making them. Other authors[54] have produced similar types of apparatus, whose general principles are illustrated in Fig. 5.16. The hopper of the usual spreader is surmounted by a container divided diagonally. Each of the compartments so formed contains a slurry of a different adsorbent; the slurries pass into the lower chamber, which contains rotating mixing blades, and the mixed slurries pass out at the bottom on to the glass plates in the usual way. The result is a plate whose composition changes regularly from 100% A at one side to 100% B at the other. Examples of gradient plates which can be made in this way would have adsorbents of two different activities, such as silica gel and kieselguhr, or of two different acidities, such as an acidic and a basic alumina (pH gradient plate).

In the most interesting use of gradient plates the solvent flows along the gradient (direction X in Fig. 5.17), so that different conditions obtain at each stage in the separation. The alternative is to make the solvent flow across the gradient (direction Y in Fig. 5.17), when each spot moves in

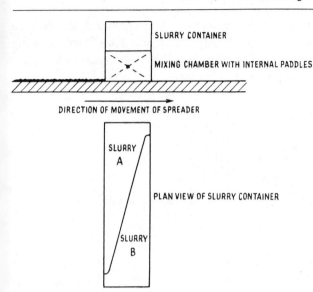

Fig. 5.16. Gradient spreader

a constant environment different from that of each of the others. This may be a useful way of finding the best mixed adsorbent for a particular separation, but the same object can be achieved with a plate made in the form of bands or strips, with a more or less abrupt change from one adsorbent to the

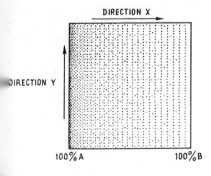

Fig. 5.17. Gradient plate

next. A 'band' spreader, which is much simpler than a gradient spreader, can easily be made from any 'moving spreader' apparatus[55] (Fig. 5.18), by the insertion of metal or plastics partitions in the hopper.

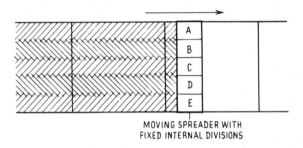

MOVING SPREADER WITH
FIXED INTERNAL DIVISIONS

Fig. 5.18. Band spreader

Modified glass plates. A very simple way of making chromatographic plates is to use reeded glass as the support. The slurry can be spread with a ruler, resting on the ridges, and the result is a number of strips parallel to each other. One spot is placed on each. Minor irregularities may give erratic R_f values, and the method is most suitable for qualitative separations. Two possibilities, with reeded glass of different designs, are illustrated in Fig. 5.19.

Wedge layers. Plates can be made which vary in thickness from side to side (wedge layers), by adapting the adjustable edge of a spreader. An improved separation of herbicides on a silica gel/kieselguhr layer has been reported[56], on plates

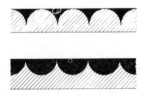

Fig. 5.19. Plates made
on reeded glass (sections)

varying in thickness from 2000 μm (at the start) to 100 μm.
Grooved plates. The use of a plate with a groove at the origin
has been described[71]. The plate (which is available commercially) has a rounded groove about 13 mm wide and
1.5 mm deep ground parallel to one edge (Fig. 5.20). The
adsorbent is spread in any conventional way, giving a
considerable increase in thickness in the groove. There is thus
afforded the possibility of a much greater loading at the
origin than will be tolerated by the rest of the plate. When
the mixture has been loaded along the groove, a volatile
solvent is used to remove trace constituents to a new origin
just beyond the groove. The solvent is allowed to evaporate,
the adsorbent within and below the groove is removed, and
the chromatographic separation of the trace constituents of
the mixture is carried out in the usual way, starting from the
new origin. Grooved plates are said to be an improvement on
wedge layers and to fulfil the same purpose; that is, the
removal of large amounts of extraneous matter in order to
examine minor components of a mixture. An example would
be the removal of proteinaceous substances in biological
extracts.
Ready-made plates on flexible supports. Several types of
prepared plate on a flexible support are available commercially. Cellulose paper sheets can be obtained loaded with

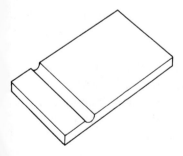

Fig. 5.20. Grooved plate

an adsorbent (about 22% SiO_2, or about 7.5% Al_2O_3, by weight). These sheets are treated like paper, except that the adsorbent can be activated to some extent before use, at a temperature not exceeding 200°C. It is said that some separations of lipids have been achieved on a silica-loaded paper which could not be achieved on silica gel on glass plates. This must have been due to adsorption or partition effects by the cellulose support.

In another commercial product[57] the adsorbent is spread on a polyester sheet. A small amount of polyvinyl acetate is incorporated, to act as a binder. The resulting material can be handled like paper. A published example of the use of these sheets is the separation of porphyrin esters by normal and reversed-phase chromatography[58].

A flexible support which is chemically more inert than those just mentioned is glass fibre paper. It can be obtained with an adsorbent included in manufacture, but it is not difficult to impregnate the sheet chemically. For impregnation with silica gel, for example, it would be dipped in potassium silicate solution, dried, dipped in hydrochloric acid, and dried again at 100°C. For impregnation with alumina, the solutions would be of aluminium chloride and ammonia, and the final drying would be at 500°C.

Ready-made flexible thin-layer plates are expensive, but they are undoubtedly convenient.

Gradient elution. The technique of gradient elution (p. 24), in which the composition of the mobile phase changes as the chromatographic development is in progress, has been applied to thin-layer chromatography. The plate stands as usual in one solvent, and the second is added slowly, with magnetic stirring. A constant level device is necessary. Examples of the use of this method are given by Wieland *et al.* [68], and by Rybicka[69] (for the separation of glycosides).

Multiple development. When a one-way chromatographic run has been completed, a better separation can often be

Substances separated	Adsorbent	Mobile phase	Method of location	Remarks	Ref. No.
amino-acids	silica gel	1. $MeOH/CHCl_3/NH_3/H_2O$ 2. $PhOH/H_2O$	ninhydrin	2-way	7
	alumina	$BuOH/EtOH/H_2O$	ninhydrin		59
	cellulose	1. $MeOH/CHCl_3/NH_3/H_2O$ 2. $MeOH/py/H_2O$	ninhydrin	2-way; no binder	29
	starch	1. $BuOH/Me_2CO/Me_2NH/H_2O$ 2. $isoPrOH/HCO_2H/H_2O$ or $isoPrOH/AcOH/py/H_2O$	ninhydrin	2-way	35
fatty acids	silica gel	1. $MeCN/Me_2CO/C_{12}H_{26}$ 2. $Pr_2O/n\text{-}C_6H_{14}$	pH indicator or 50% H_2SO_4 iodine diphenyl-carbazonc	2-way; adsorbent with binder and $AgNO_3$	9
unsaturated fatty acids	silica gel	light pet. /Et_2O	conc.H_2SO_4 iodine		62
lipids	silica gel	pet./Et_2O/AcOH	conc.H_2SO_4		10
hydrocarbon oils synthetic ester oils	silica gel	$CHCl_3/C_6H_6$	fluorescence conc.H_2SO_4		61
sterols	silica gel	$CHCl_3/Me_2CO$	fluorescence	$AgNO_3$ included in layer	60
simple sugars	silica gel	$BuOH/AcOH/Et_2O/H_2O$	conc.H_2SO_4 $KMnO_4$		63
phenols	silica gel	$C_6H_6/MeOH$	diazotised benzidine	layer impregnated with formamide	64
2,4-dinitro-phenylhydrazones	kieselguhr	light petroleum	coloured	plate impregnated with phenoxyethanol	65
alkali metals	silica gel	EtOH/AcOH	violuric acid	multiple development	66

obtained by a second development in the same direction with the same, or a different, solvent. Up to four developments of one plate have been recommended, for example, by Urbach[65] for the separation of aliphatic 2,4-dinitrophenyl-hydrazones.

Reversed phase partition chromatography. If the adsorbent is coated with a non-polar medium, such as a hydrocarbon (for example, undecane) or a silicone, reversed phase partition chromatography can be carried out, in the same way as with impregnated papers (p. 157). The impregnation can be done by very cautious dipping of the plate in a solution of the non-polar medium in a volatile solvent, but it is easier to allow the solution to flow through the adsorbent in the usual chromatographic manner, and then to allow the solvent to evaporate before the samples are applied.

Uses of thin-layer chromatography

It has already been pointed out that thin-layer chromatography has been used in very many fields. A few representative examples are given in Table 5.2.

It will be noticed that the examples include both hydrophilic and hydrophobic substances, and several adsorbents, although of these silica gel appears most often, because it is the most popular and useful. For a more detailed list of applications, one of the monographs on thin-layer chromatography should be consulted, which are listed in the bibliography on p. 373. Reviews of special fields have been published by Padley[8] (lipids), Pataki[67] (amino-acids and peptides), and Rettie and Haynes[70] (dyes).

REFERENCES

1. Meinhard, J. E., Hall, N. F., *Analyt. Chem.*, 1949, **21**, 185.
2. Kirchner, J. G., Miller, J. M., Keller, J. G., *ibid.*, 1951, **23**, 420.
3. Stahl, E., *Thin-layer Chromatography*, Academic Press, London, 1965

4. Stahl, E., *Thin-layer Chromatography*, Editor Martini-Bettolo, G.B., Elsevier, London, 1964, p. 1.
5. Brenner, M., Niederwieser, A., *Experientia*, 1960, **16**, 378.
6. Waldi, D., Schnackerz, K., Munter, F., *J. Chromatog.*, 1961, **6**, 61.
7. Stahl, E., Kaltenbach, V., *ibid.*, 1961, **5**, 351.
8. Padley, F. B., *Thin-layer Chromatography*, Editor Martini-Bettolo, G.B., Elsevier, London, 1964, p. 87.
9. Bergelson, L. D., Dyatlovitskaya, E. V., Voronkova, W. V., *J. Chromatog.*, 1964, **15**, 191.
10. Malins, D. C., Mangold, H. K., *J. Amer. Oil Chemists' Soc.*, 1960, **37**, 383 & 576.
11. Stahl, E., Trennheuser, L., *Arch. Pharm.*, 1960, **293/65**, 826.
12. Seiler, H., *Helv. Chim. Acta*, 1961, **44**, 1753; 1962, **45**, 381.
13. Bennett, R. D., Heftmann, E., *J. Chromatog.*, 1962, **9**, 353 & 359.
14. Neher, R., *Steroid Chromatography*, Elsevier, Amsterdam, 1964.
15. Tschesche, R., Lampert, F., Snatzke, G., *J. Chromatog.*, 1961, **5**, 217.
16. Mottier, M., Potterat, M., *Analyt. Chim. Acta*, 1955, **13**, 46.
17. Sarsunova, M., Schwarz, V., *Pharmazie*, 1962, **17**, 527.
18. Cerny, V., Joska, J., Labler, L., *Coll. Czech. Chem. Comm.*, 1961, **26**, 1658.
19. Blattna, J., Davidek, J., *Experientia*, 1961, **17**, 474.
20. Davidek, J., Blattna, J., *J. Chromatog.*, 1962, **7**, 204.
21. Weill, C. E., Hanke, P., *Analyt. Chem.*, 1962, **34**, 1736.
22. Knappe, E., Peteri, D., *Z. analyt. chem.*, 1962, **188**, 184 & 352.
23. Kaufmann, H. P., Makus, Z., Khoe, T. H., *Fette u. Siefen*, 1961, **63**, 689.
24. *idem*, *ibid.*, 1962, **64**, 1.
25. Honegger, C. G., *Helv. Chim. Acta*, 1961, **44**, 173.
26. Peereboom, J. W. C., Beekes, H. W., *J. Chromatog.*, 1962, **9**, 316.
27. Vaedlke, J., Gajewska, A., *ibid.*, 1962, **9**, 345.
28. von Arx, E., Neher, R., *ibid.*, 1963, **12**, 329.
29. Bujard, El., Mauron, J., *ibid.*, 1966, **21**, 19.
30. Wollenweber, P., *ibid.*, 1962, **7**, 557.
31. Teichert, K., Mutschler, E., Rodelmeyer, H., *Z. Analyt. Chem.*, 1961, **181**, 325.
32. Randerath, K., Struck, H., *J. Chromatog.*, 1961, **6**, 365.
33. Randerath, K., *Angew. Chem.*, 1961, **73**, 436.

34. Berger, J. A., Meyniel, G., Petit, J., *Compt. Rend.*, 1962, **255**, 1116.
35. Petrovic, S. M., Petrovic, S. E., *J. Chromatog.*, 1966, **21**, 313.
36. Birkhofer, L., Kaiser, C., Meyer-Stoll, H. A., Suppan, F., *Z. Naturforsch*, 1962, **17b**, 352.
37. Davidek, J., *J. Chromatog.*, 1962, **9**, 363.
38. Davidek, J., Davidkova, E., *Pharmazie*, 1961, **16**, 352.
39. Hofmann, A. F., *Biochim. Biophys. Acta*, 1962, **60**, 458.
40. Johansson, B. G., Rymo, L., *Acta Chem. Scand.*, 1962, **16**, 2067.
41. *idem, ibid.*, 1964, **18**, 217.
42. Morris, C. J. O. R., *J. Chromatog.*, 1964, **16**, 167.
43. van Damm, N. J. D., Maas, S. P. J., *Chem. & Ind.*, 1964, 1192.
44. Lees, J. M., DeMuria, P. J., *J. Chromatog.*, 1962, **8**, 108.
45. Bekersky, I., *Analyt. Chem.*, 1963, **35**, 261.
46. Morita, K., Harata, F., *J. Chromatog.*, 1963, **12**, 412.
47. Morgan, M. E., *ibid.*, 1962, **9**, 379.
48. Curtis, P. J., *Chem. & Ind.*, 1966, 247.
49. Litt, G. L., Johl, R. G., *J. Chromatog.*, 1965, **20**, 605.
50. Pataki, G., Kelemen, J., *ibid.*, 1963, **11**, 50.
51. Shellard, E. J., *Lab. Prac.*, 1964, **13**, 290.
52. Dallas, M. S. J., *J. Chromatog.*, 1965, **11**, 267.
53. Stahl, E., *Angew. Chem.*, 1964, **3**, 784.
54. Warren, B., *J. Chromatog.*, 1965, **20**, 603.
55. Abbott, D. C., Thompson, J., *Chem. & Ind.*, 1965, 310.
56. *idem, ibid.*, 1964, 481.
57. Przybylowicz, E. P., Standenmeyer, W. J., Perry, E. S., Baitsholtz, A. D., *J. Chromatog.*, 1965, **20**, 506.
58. Chu, T. C., Chu, E. J., *ibid.*, 1966, **21**, 47.
59. Mottier, M., *Mitt. Gebiete Lebensm. u. Hyh.*, 1958, **49**, 454; 1956, **47**, 372.
60. Dillullio, N. W., Jacobs, C. S., Holmes, W. L., *J. Chromatog.*, 1965, **20**, 354.
61. Crump, G. B., *Nature*, 1962, **193**, 674.
62. Mangold, H. K., Kammereck, R., *Chem. & Ind.*, 1961, 1032.
63. Hay, G. W., Lewis, B. A., Smith, F., *J. Chromatog.*, 1963, **11**, 479.
64. Waksmundzki, A., Manko, R., *Stationary Phase in Paper and Thin-layer Chromatography*, Editors Macek, K., Hais, I. M., Elsevier, London, 1965, p. 221.

65. Urbach, G., *J. Chromatog.*, 1963, **12**, 196.
66. Seiler, H., Rothweiler, W., *Helv. Chim. Acta*, 1961, **44**, 941.
67. Pataki, G., *Dünnschichtchromatographie in der Aminosäure und Peptid-Chemie*, de Gruyter, Berlin, 1966.
68. Wieland, T., Lüben, B., Determan, H., *Experientia*, 1962, **18**, 430.
69. Rybicka, S. M., *Chem. & Ind.*, 1962, 308.
70. Rettie, G. H., Haynes, C. G., *J. Soc. Dyers & Colourists*, 1964, **80**, 629.
71. Collins, R. F., *Chem. & Ind.*, 1969, 614.

Chapter Six

Model Experiments in Chromatographic Techniques

The experiments described in this chapter are a few of those which have been found by the authors to be useful in teaching the fundamental techniques of chromatography to undergraduate students, and to members of short courses in Practical Chromatography. The experiments have been chosen to give an example of each of the main procedures, and all have been found to work well in the hands of the inexperienced. No particular originality is claimed, although it is not always possible to quote original sources, because several published methods may have been blended, and then further adapted, by the present authors or their colleagues. In many cases the experiments are designed for use with a particular material or apparatus, often merely because that happened to be available. It is usually not difficult to modify the methods so that other similar equipment can be used. All the experimental figures quoted have been determined by the authors (or their students) for the conditions described. This includes the R_f values for paper and thin-layer experiments, which show some slight differences from the figures published elsewhere.

Paper experiments are easy to perform, and there are many variations in technique (ascending, descending, horizontal, reversed phase, etc.), with quite simple apparatus. For those reasons a comparatively large number of such experiments has been included. They are intended to illustrate the main types of separation which are possible on paper, as well as to give practice in the different techniques.

Some experiments in thin-layer chromatography are included, to illustrate some applications of the technique. Specific methods of preparing plates have, however, been omitted, because most commercial suppliers provide adequate instruction sheets for the use of their equipment, and 'do-it-yourself' methods are described in Chapter 5. The experiments are listed in Table 6.1.

Table 6.1   *List of model experiments*

I.	Column Chromatography Experiments 1—3, Adsorption Experiment 4, Partition
II.	Ion-exchange Chromatography Experiments 5 and 6
III.	Paper Chromatography — Inorganic Experiments 7—11
IV.	Paper Chromatography — Organic Experiments 12—15
V.	Electrophoresis Experiment 16
VI.	Gas Chromatography Experiments 17 and 18
VII.	Thin-layer Chromatography Experiments 19—22

I   Column Chromatography

Experiment 1: *Adsorption — stepwise elution*

Separation of cis- and trans-azobenzene [1, 2]

Ordinary samples of azobenzene contain about 1% of the *cis*-isomer; irradiation with ultra-violet light increases this proportion. Dissolve 0.25 g of azobenzene in 25 cm^3 of light petroleum (b.p. 60—80°C), put the solution in a loosely stoppered test tube and stand it under an ultra-violet lamp for about 30 minutes.

Prepare a column 1.5 cm in diameter and 10 cm long (p. 20), using light petroleum (b.p. 60–80°C) as the solvent and alumina* (about 25 g) as the adsorbent. Allow the solvent to run down until it is just above the top of the column, and cautiously add the azobenzene solution, letting the solvent flow out at a rate of 1–2 cm³ per minute. Elute with light petroleum, using the same flowrate. The *cis*-azobenzene remains as a compact band at the top, and a more diffuse band of *trans*-azobenzene moves down the column. Change the receiver when this band is just about to leave the column, and collect the yellow solution of the *trans*-isomer. Evaporation of the solvent gives *trans*-azobenzene, m.p. 68°C.

When the eluate becomes colourless change the eluting solvent to light petroleum containing 1% of methanol to elute the *cis*-azobenzene. Collect the fraction containing the *cis*-isomer, wash with water to remove the methanol, dry over sodium sulphate and evaporate the solvent under reduced pressure to obtain *cis*-azobenzene, m.p. 61°C.

For a more elaborate version of this experiment, using a spectrophotometric method to examine the separated isomers, see Wilson *et al*[3].

Experiment 2: *Adsorption – stepwise elution*

Separation of permanganate and dichromate

The speed of separation makes this experiment very useful as a lecture demonstration, and the simple quantitative finish makes it suitable for students at an elementary level.

Make up a column of alumina (see Experiment 1) 1.5–2.0 cm wide and about 15 cm long (30–40 g), using 0.5 mol dm⁻³ nitric acid as the solvent. Wash with about 25 cm³ of the same acid at a rate of 2 cm³ per minute.

---

* The alumina used for the experiment as described was BDH 'Alumina for Chromatography'. Other aluminas could be used, but not necessarily with the results described here.

The solution to be separated can be made by mixing equal volumes of 0.02 mol dm^{-3} potassium permanganate and 0.016 mol dm^{-3} potassium dichromate. Carefully introduce 10 cm^3 of the solution onto the column, and follow it with 0.5 mol dm^{-3} nitric acid, maintaining the rate of flow at 2 cm^3 per minute.

The dichromate is adsorbed at the top of the column, and the permanganate passes through as a rather diffuse band. When this band is near the bottom, start collecting the eluate in a conical flask. Remove the flask when all the permanganate has passed through. Then change the eluting solution to 1.0 mol dm^{-3} sulphuric acid, to elute the dichromate, and collect the solution in a second conical flask.

To determine the concentration of permanganate, add a known excess of standard hydrogen peroxide, and back-titrate with standard 0.02 mol dm^{-3} potassium permanganate. To determine the dichromate, add a known excess of standard iron(II) ammonium sulphate, and back-titrate with standard 0.02 mol dm^{-3} potassium permanganate.

Experiment 3: *Adsorption – frontal analysis*

Separation of chloride from sulphate

Prepare a column of alumina as described in Experiment 2, using 0.5 mol dm^{-3} nitric acid as the solvent. Wash with 0.5 mol dm^{-3} nitric acid, and test that the washings are free from chloride and sulphate.

The solution to be separated is a mixture of hydrochloric and sulphuric acids (about 1.0 mol dm^{-3} HCl and 0.5 mol dm^{-3} H$_2$SO$_4$). Make up 150 cm^3, add it slowly to the top of the column, and pass it through continuously, at a flowrate of 2 cm^3 per minute.

As soon as the solution has been added to the column, start collecting 2 cm^3 fractions in a series of test tubes. Divide each fraction into two parts, and test one part for chloride (silver nitrate), and the other for sulphate (barium chloride). Record (*a*), the volume collected before chloride ion first appears, and (*b*), the volume which contains only chloride, before sulphate first appears. (*a*) should be 30–40 cm^3, and (*b*) 14–20 cm^3; the exact values depend largely on the nature of the alumina.

Experiment 4: *Partition – elution analysis*

Separation of cobalt and nickel[4]

Although metals can be estimated quantitatively by separation on paper sheets, it is often preferable to use a cellulose column if the evaluation is to be by elution of the substance followed by a normal quantitative finish. This experiment illustrates the separation of cobalt from nickel by simple elution from a column of cellulose powder.

Prepare a column of cellulose powder (Whatman fibrous CfO) about 10 cm long and 1.5–2.0 cm in diameter (p. 39), using as solvent acetone/conc. hydrochloric acid 98 : 2 v/v. Wash the column by allowing about 50 cm^3 of solvent to flow through slowly. The solution for analysis should contain about 2.0 g dm^{-3} of each metal, and should be made up by dissolving the chlorides in water, with the minimum amount of hydrochloric acid.

Take 5 cm^3 of the solution (quantitatively) in a 100 cm^3 beaker, evaporate almost to dryness and take up the residue in 2 cm^3 of hydrochloric acid (80 cm^3 concentrated acid and 20 cm^3 water). Add about 1.5 g of cellulose powder, which soaks up the solution to make a friable mass. Transfer to the top of the column, and add 5 cm^3 of solvent. Stir up and then gently press down to form a homogeneous compact column. Allow solvent to flow at about 2 cm^3 per minute (wash out the beaker with solvent and add the washings to the column). The cobalt is eluted as a green solution of a complex ion, and the nickel is retained on the column as a nearly colourless band.

Collect about 100 cm^3 of the eluate; that is, until all the green solution has passed through. For a quantitative result for cobalt, evaporate the acetone from the eluate on a water bath, add 2 cm^3 conc. nitric acid and 1 cm^3 conc. sulphuric acid, and heat to SO_3 fumes. Cool, dilute with a few cm^3 of water, and determine cobalt. A colorimetric method is the most convenient.

To show that the nickel is retained on the column, pour in a few cm^3 of 2 mol dm^{-3} ammonia, and then dimethylglyoxime solution. The nickel will appear as a pink band near the top of the column.

II   Ion-Exchange Chromatography

Experiment 5: *Ion-exchange column – stepwise elution*

Separation of iron(III) *and copper*(II)[5]

The separation involves the stepwise elution of iron(III) by phosphoric acid and then of copper(II) by hydrochloric acid, from a strongly acidic cation-exchange resin such as Zeo-Karb 225, Amberlite IR-120 or Dowex 50. If hydrochloric acid alone is used the order of removal of the ions from the column is reversed, and a poorer separation is obtained (p. 54).

Make up a column about 1.5 cm in diameter and 10–15 cm long (p. 52), using Zeo-Karb 225, mesh size 52–100 (8% DVB), (supplied in the sodium form). Slurry the resin with water, allow to settle, and decant the supernate. Repeat until the washings are clear, and then pack the column. Run about 200 cm^3 of 5 mol dm^{-3} hydrochloric acid through the column, and then water until the eluate is no longer acid. The column is then in the hydrogen form.

The solution containing iron and copper is about 0.05 mol dm^{-3} with respect to each metal, made up from the chlorides. Take about 20 cm^3 of the solution and run it carefully onto the column. Wash with 100 cm^3 of water to remove the liberated acid, and then elute the iron with 10% phosphoric acid (15% syrupy phosphoric acid in water by volume), using a flowrate of about 2 cm^3 per minute. Collect 2 cm^3 fractions, and, since the phosphate complex is colourless, test each fraction with potassium cyanoferrate(II).

When no more iron can be detected change the eluting agent to hydrochloric acid (equal volumes of concentrated acid and water); the copper is rapidly removed from the column. Collect fractions as before, and test for copper with the same reagent – K$_4$[Fe(CN)$_6$].

Note (*a*) the volume eluted before iron first appears; (*b*) the volume which contains iron; (*c*) the volume collected after all the iron has been eluted, and before copper first appears; and (*d*) the volume which contains copper. Average values are: (*a*) 15 cm^3; (*b*) 100 cm^3; (*c*) 20 cm^3; and (*d*) 10 cm^3.

The separated metals can be estimated by normal methods of analysis (for example with a 'Spekker'), if the original solution is measured quantitatively.

Experiment 6: *Ion-exchange column – stepwise elution*

Separation of halides [6]

A quantitative separation of chloride, bromide, and iodide can be achieved on a strongly basic anion-exchange resin such as De-Acidite FF, Amberlite IRA-400, Dowex 1, or Dowex 2, in the nitrate form $RNMe_3^+ \cdot NO_3^-$. When the halide solution is introduced, the halide ions displace nitrate ions to form a narrow band at the top of the column. Elution with a solution containing nitrate ions (sodium nitrate) brings about displacement of chloride rapidly, bromide less rapidly, and iodide very slowly. The elution is followed by titration with silver nitrate.

Stir some De-Acidite FFIP, mesh size 52–100, 3–5% average cross-linking, with water, allow to settle and decant the liquid. Repeat until the supernate is clear. Make up a column about 1.5 cm in diameter and 10–15 cm long. Wash slowly with 0.5 mol dm^{-3} sodium nitrate until the washings are free from chloride (the resin being supplied in the chloride form).

The solution to be separated contains chloride and iodide (sodium or potassium salts), and is about 0.5 mol dm^{-3} in each. Introduce 2 cm^3 of the solution on to the column, and elute with 0.5 mol dm^{-3} sodium nitrate at 2 cm^3 per minute. Collect the eluate in 10 cm^3 fractions, using a measuring cylinder. Decant each fraction into a conical flask and titrate with standard 0.03 mol dm^{-3} silver nitrate, using potassium chromate as the indicator. Do a blank on 10 cm^3 of the sodium nitrate solution.

When the titre falls to, or close to, zero, start eluting with 3 mol dm^{-3} sodium nitrate, which will accelerate the displacement of iodide ion. Titrate each 10 cm^3 fraction as before with standard silver nitrate, but use Rose Bengal or dimethyl di-iodofluorescein as the indicator (use the minimum amount, as otherwise the colour change is hard to see when only small amounts of iodide are present). Continue until all the iodide has been eluted.

Plot a graph of halide concentration (silver nitrate titres) against volume of effluent. Summation of the titres will lead to the concentration of chloride and iodide in the original solution. A typical curve is shown in Fig. 6.1. If bromide is present the concentrations of sodium nitrate used for elution are: for chloride, 0.3 mol dm^{-3}; for bromide, 0.6 mol dm^{-3}; for iodide, 3 mol dm^{-3}. The presence of bromide as well as the other halides necessitates a somewhat larger column containing about 30 g of resin.

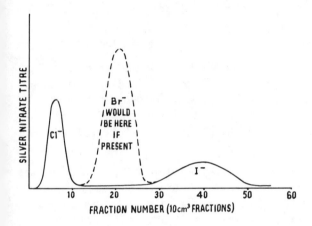

Fig. 6.1. Separation of halides (Experiment 6)

III Paper Chromatography — Inorganic

Experiment 7: *Ascending and horizontal methods*

Separation of cobalt, manganese, nickel, and zinc

This experiment illustrates the use of gas jars for ascending separations, and of the horizontal method. It also affords a comparison between three different locating reagents, and different papers.

Solution: mixture of the chlorides of the four metals (about 0.05 g of each metal in 100 cm^3).

Solvent: acetone/conc. HCl/water 87 : 8 : 5 v/v.

Paper: *   No. 1, No. 3 MM, (reels 3 cm wide, or cut from sheets).
             No. 1, No. 2, and No. 4 (12.5 cm circles).
Locating   (*a*) rubeanic acid/salicylaldoxime/alizarin (RSA).
  reagents: (*b*) diphenylcarbazide.
             (*c*) sodium pentacyanoammine ferrate(II)/rubeanic acid
             (PCFR).
Tanks: (*a*) gas jars 7 cm x 30 cm, with suspension as shown in
             Fig. 3.7 (*b*).
             (*b*) Petri dishes 11 cm in diameter.

Ascending method

Cut three strips of No. 1 paper 24 cm long. Draw a transverse line 2 cm
from one end and punch a small hole in the other. Apply the test
solution to the middle of the line with a platinum loop about 3 mm
wide (p. 145), so that a spot about 0.5 cm in diameter is formed. Two
drops should be applied, giving a total volume of about 10 mm^3, and
thus about 2.5 μg of each metal.

Put solvent in the gas jars to a depth of about 3 cm, and roll the jars
round in a nearly horizontal attitude to wet the walls with solvent.
Raise the hooks to the highest position, attach the strips, and replace
the lids on the gas jars. Allow to stand for a few moments, and then
lower the strip so that the end is 1 cm below the liquid surface. A run
of 12–15 cm is sufficient; when the solvent has risen to that height
remove the strips from the jars and hang to dry (mark the solvent
front while it is still visible). Spray one strip with each of the reagents
(*a*), (*b*), and (*c*); (see Appendix). Repeat the experiment with No. 3 MM
paper.

Horizontal method

Method (*i*) – Fig. 3.12. Locate the centre of a 12.5 cm paper disc, and
form a wick by making two parallel cuts 4 mm apart as shown in Fig.
3.11 (p. 119). (A simple template to locate the centre can be made by
taking a paper circle, folding into quarters as for ordinary filtration, and
cutting off the extreme tip; when the paper is opened out there will be

* In all the paper chromatography and electrophoresis experiments
described, the code numbers given are those of Whatman papers.

a small hole at the centre). Draw a line at a radius of 0.5–1.0 cm from the centre and apply the solution with a capillary tube as a streak following this line (p. 146); about 1.5 cm in a melting-point tube will be required. Allow to dry. Put some solvent in a Petri dish so that the dish is about half full, place the paper on top so that the wick dips in the solvent (a little may have to be trimmed off the end of the wick for neatness), and put a second dish on top. Allow to run until the solvent front reaches the edge of the dish, and then remove and dry, and spray with one of the locating reagents.

Method (ii) – Fig. 3.13. Locate the centre of the paper disc, and make a hole with a No. 1 cork borer. Make a small wick with a piece of the same type of paper rolled into a tube, and long enough to reach to the bottom of a Petri dish (p. 121). The exact amount of paper necessary for the wick has to be found by trial and error, but no part of it should be of more than one thickness of paper. Draw a circle of 1.5–2.0 cm diameter and on it apply the samples with a platinum loop at about 1.5 cm intervals. Allow to dry, put the paper on a Petri dish containing solvent, cover with a second dish, and run, dry, and spray as before.

The horizontal methods should be tried with Nos. 1, 2, and 4 papers.

The order of the metals from the start, and the approximate R_f values for the ascending method are:

$$\text{Ni, } 0.05; \quad \text{Mn, } 0.30; \quad \text{Co, } 0.50; \quad \text{Zn, } 0.90$$

The visible solvent front is the 'dry' front. The 'wet' front has an R_f value of about 0.75, and a pale yellow band may be seen at this point (before application of the locating reagent) due to traces of iron (Fig. 6.2). It should be noted that the iron (very soluble in the wet solvent) and the zinc (very soluble in the dry solvent) give rather diffuse zones, whereas the other metals give much more compact spots (see the Appendix for details of the reagents and colours).

Experiment 8: *Descending method*

Separation of anions (halides) [7]

This experiment illustrates the descending technique as used for inorganic substances, and also shows how anions can be separated from each other.

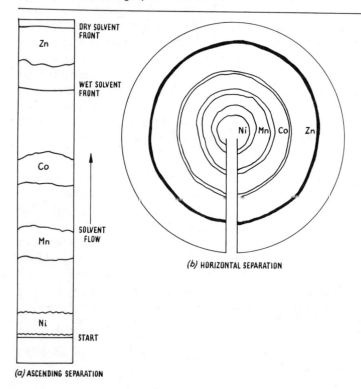

(a) ASCENDING SEPARATION

(b) HORIZONTAL SEPARATION

Fig. 6.2. Separation of metals on Whatman No. 1 paper
(Experiment 7)

Solution: mixture of sodium fluoride, chloride, bromide and iodide
             (about 1 mg of each in 100 cm³).
Solvent: pyridine/water 90/10 v/v.
Paper: No. 1 and No. 3 MM (reels 3 cm wide).
Locating   (a) silver nitrate/fluorescein.
  reagents: (b) zirconium/alizarin S.
Tank: glass cylinder type with stainless steel interior frame for
             supporting the trough.

Cut off two strips of each type of paper 45 cm long, and fold and
mark as shown in Fig. 6.3 (a). Apply a spot of the test solution to each
strip (2 drops with a platinum loop) and dry. Put a little solvent
mixture in the dish at the bottom of the tank, and then arrange the

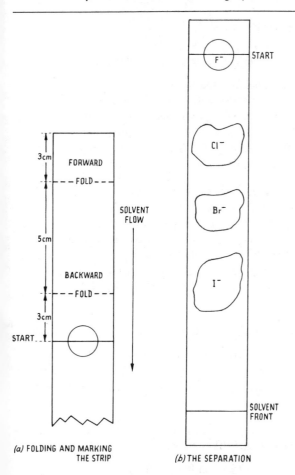

Fig. 6.3. Separation of halides (Experiment 8)

strips with their ends in the solvent trough so that the other ends hang freely. Replace the lid and carefully fill the trough with solvent through the hole in the lid by means of a pipette; replace the glass disc over the hole. No equilibration time is needed. Allow to run for 3-4 hours, and then remove and dry the strips. Treat one of each type with reagent (*a*) to detect Cl⁻, Br⁻, and I⁻, and the other with reagent (*b*) to detect F⁻, as described in the Appendix (see Fig. 6.3 (*b*)). Approximate R_f values are: F⁻, 0.00; Cl⁻, 0.24; Br⁻, 0.47; I⁻, 0.71.

Experiment 9: *Ascending method*

Rapid small-scale separation of metals [8]

This experiment illustrates the use of a very simple and rapid method of separating metals, with comparison samples chromatographed on the same sheet. The paper used is type CRL/1, No 1 and No. 4 (p. 114), and the tank is a tall form beaker, without pouring spout, closed with a clock glass (beakers having pips or minor irregularities on the rim should be discarded).

Apply the reference and test solutions with a 3 mm platinum loop (p. 145) as shown in Fig. 3.6 (*a*). When the spots are dry form the paper into a cylinder and fasten with a paper clip at the top (Fig. 3.6 (*b*)). Put solvent in the beaker to a depth of 0.5 cm, insert the paper and replace the clock glass. When the solvent reaches the top of the slits, remove the paper, stand it to dry on the clock glass, and then open it out and apply the locating reagent. Do each separation with No. 1 and No. 4 paper.

(*a*) *Separation of cobalt, copper, iron, manganese, nickel, and zinc*

Reference solutions of the chloride of each metal, and of 'unknowns' with two or more cations are required. The concentration of each metal should be about 5 mg cm^{-3}. Label each strip and put the spots on in some such order as:

(1) unknown; (2) Cu; (3) Co; (4) unknown; (5) Fe; (6) Mn; (7) unknown; (8) Ni; (9) Zn; (10) unknown.

Solvent: acetone/conc. HCl/water 87 : 8 : 5 v/v.
*Locating
reagent:*   RSA or PCFR (see Experiment 7).

(*b*) *Separation of analytical Group IIA metals*

Solutions required are of chlorides of the same concentrations as in separation (*a*). Put the spots in some such order as:

(1) unknown; (2) Bi; (3) Cd; (4) unknown; (5) Cu; (6) Fe; (7) unknown; (8) Hg; (9) Pb; (10) unknown.

Solvent: butan-1-ol saturated with 3 mol dm^{-3} hydrochloric acid.
Locating either spray with diphenylcarbazide, or dip in yellow
reagent: ammonium sulphide.

The iron solution is included because traces of that metal usually appear in the chromatograms of the others, having been present as an impurity in the acid or the paper.

Experiment 10: *Ascending method*

Quantitative separation of cobalt, copper, and nickel[8]

This experiment is a simple quantitative separation, the final evaluation being done by visual comparison of the coloured spots with standards. An accuracy of ±5% is possible if the procedure is carefully followed, and the standards are accurately made up.

Solutions: (*a*) Standard solutions:
S1–4.0 μg; S2–2.0 μg; S3–1.0 μg; S4–0.50 μg;
S5–0.25 μg of each metal per 0.01 cm^3
Solution S1 is made up of the following:
$CuCl_2 \cdot 2H_2O$–282 mg ⎫ made up to 250 cm^3 in water
$CoCl_2 \cdot 6H_2O$–395 mg ⎬ with the minimum amount of
$NiCl_2 \cdot 6H_2O$–395 mg ⎭ hydrochloric acid.
S2–S5 are made by quantitative dilution of S1.
(*b*) Solutions containing an unknown amount of each metal; the concentrations should fall within the limits of S1 and S5.
Solvent: butanone/conc. HCl/water 75 : 15 : 10 v/v.
Paper: CRL/1, No. 1.
Locating
reagent: rubeanic acid.
Tank: 1 dm^3 beaker and clock glass, as in Experiment 9.

Make up some solvent mixture and put 25 cm^3 into the 1 dm^3 beaker. Roll the solvent round the walls and cover with the clock glass. Put about 2 cm^3 of each standard solution and one unknown into labelled test tubes, to each add 0.5 g of potassium hydrogen sulphate, warm, and then cool to room temperature.

Apply 0.01 cm^3 quantitatively to each strip with a pipette (p. 145), in the order S1, U, S2, U ... (U is the unknown). Label each strip. Make the sheet into a cylinder by means of a paper clip at the top, and put it in a 600 cm^3 beaker floating in a boiling water-bath. Leave it for three minutes to dry, and then immediately put it into the beaker containing the solvent. Replace the cover, and allow to run until the solvent front is just above the top of the slots (about 50 min).

Remove the sheet, allow to dry in the air for five minutes, and then stand it in an atmosphere of ammonia for two minutes (a convenient way is to put it in a covered 600 cm^3 beaker, in the bottom of which is a 25 cm^3 beaker containing 0.880 ammonia). Immediately open out the cylinder and spray the paper evenly on both sides with rubeanic acid solution; dry and estimate the concentration of the unknown solution by visual comparison of the spots.

Experiment 11: *Ascending method – ion-exchange paper*

Separation of copper, iron, and nickel[9]

This experiment illustrates the use of ion-exchange papers for the separation of metals. A strong and a weak acid cation-exchange modified cellulose paper are used, in the form in which they are supplied. The 'dry' technique is therefore adopted (p. 159). The method is similar to that of Experiment 7.

Solution: iron, copper, and nickel, as chlorides (about 2 mg of each metal per cm^3).

Eluting solution (buffer): 1.0 mol dm^{-3} MgCl$_2$ · 6H$_2$O.

Paper: (*a*) cellulose phosphate (P81) in the monoammonium form.
(*b*) carboxymethylcellulose (CM82) in the sodium form.
Supplied as sheets in each case – cut strips as required.

Locating reagent: PCFR.

Tanks: gas jars as used in Experiment 7.

Prepare two tanks, and one strip 3 cm wide of each paper as described for Experiment 7. Apply the solution as a streak (p. 145) with a capillary tube (about 10 mm^3 −1.5 cm in a melting point tube); there is no need to dry the spot. Put the paper in the tank and start the run immediately. For comparison, run a strip in a third gas jar using No. 1 paper, and the solvent used in Experiment 7. Allow a run of 10−15 cm, and then remove the sheets, dry, and apply the locating reagent.

Notice that the strong acid paper gives a better separation than the weak, and that the order of the spots is the opposite on the ion-exchange paper to that in normal chromatography (Fig. 6.4). Notice also that on P81 paper the iron does not move, because of the high stability of the iron-phosphate complex.

The approximate R_f values are given in Table 6.2.

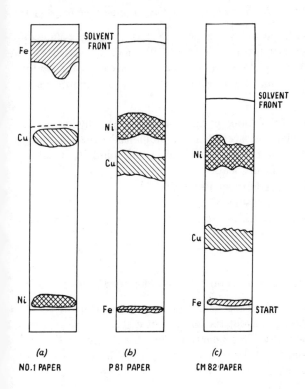

Fig. 6.4. Separation of metals on Whatman papers (Experiment 11)

Table 6.2 R_f *values for Experiment 11*

Paper	P81	CM82	No. 1
iron(III)	0.00	0.06	0.75
copper(II)	0.60	0.27	0.65
nickel(II)	0.75	0.75	0.05

IV   Paper Chromatography — Organic

Experiment 12: *Ascending method – reversed phase*

Separation of the 2,4-dinitrophenylhydrazones of lower aldehydes [10]

Aldehyde dinitrophenylhydrazones are too sparingly soluble in water for the normal method to be suitable, and a reversed phase system is therefore required (p. 157). The stationary phase used in this case is olive oil. No locating reagent is needed, as the substances are coloured.

Solutions: reference solutions of the 2,4-dinitrophenylhydrazones of formaldehyde, acetaldehyde, propionaldehyde and butyraldehyde in ethanol (1 mg cm^{-3}), and solutions containing two or more of them.

Solvent: methyl acetate/water 10 : 7 v/v.

Paper: No. 1 – sheets 46 x 57 cm.

Tank: a rectangular glass vessel 15 cm x 15 cm, and 30 cm deep, covered with a glass plate.

Reversed phase medium: 40% v/v solution of olive oil in carbon tetrachloride.

Line the tank with sheets of No. 1 paper soaked in the solvent mixture, and leave for about six hours. Cut a sheet of paper 30 cm square, and draw a pencil line 3 cm from one edge at right angles to the machine direction, which is parallel to the original longer edge of the sheet (p. 127). Mark and label seven points on this line 4 cm apart and not less than 3 cm from the end.

Dilute 50 cm^3 of the stock olive oil solution to 400 cm^3 with carbon tetrachloride. Put the solution in a 50 cm dipping tray and pass the prepared paper slowly through it with a see-saw motion, holding opposite edges with large 'bull-dog' clips. One pass is sufficient. Hang up to dry in the air (about 15 minutes).

Apply spots of each reference solution, and three of the mixture, each of about 10 mm^3. Two drops with a platinum loop about 4 mm in diameter will suffice (p. 145). Allow to dry, and then form the sheet into a cylinder with wire paper clips at top and bottom (see Fig. 3.5 (*b*)). Put 100 cm^3 of solvent mixture in the tank, insert the cylinder to stand with the spots at the lower end, and firmly replace the lid.

A run of about 16 hours is necessary (that is, overnight); after that period, remove the cylinder, mark the position of the solvent front (if it is visible), and allow to dry in the air. Remove the clips and draw circles round the yellow spots. Examine under ultra-violet light. Measure R_f values, or (if no solvent front was visible) R_{form} values, and identify the constituents of the mixture. Approximate R_f and R_{form} values are given in Table 6.3.

Table 6.3 R_f *and* R_{form} *values for Experiment 12*

	R_f	R_{form}
formaldehyde 2,4-dinitrophenylhydrazone	0.65	1.00
acetaldehyde 2,4-dinitrophenylhydrazone	0.59	0.91
propionaldehyde 2,4-dinitrophenylhydrazone	0.49	0.75
butyraldehyde 2,4-dinitrophenylhydrazone	0.35	0.54

Notice that since this is a reversed phase system the substances of lowest relative molecular mass (most soluble in polar solvents) move farthest.

Experiment 13: *Ascending method – two-way*

Separation of amino-acids [11, 12, 13]

The classical paper separation of amino-acids is the two-way method, and a number of students' experiments of that type have been published. If only the chromatographic technique is to be taught, it has seemed to the authors best to start with 'synthetic' mixtures of pure amino-acids, rather than with protein hydrolysates, with the added complications, such as desalting, which they require; thus more 'realistic' separations are left until after the basic technique has been acquired. The ascending frame method of Dent and of Smith is used in this experiment (p. 116), with Smith's multiple dipping procedure.

Solutions: reference solutions of individual amino-acids (in water, dilute hydrochloric acid, or 10% propan-2-ol), and solutions of mixtures. Suitable mixtures are listed below; each acid should be present in a concentration of about 0.25% w/v.

Solvents: 1. butan-1-ol/acetic acid/water 12 : 3 : 5 v/v.
2. phenol/water 500 g : 125 cm^3. This solvent is made up by adding 125 cm^3 of water to a normal 500 g bottle of phenol and leaving to stand in a warm place until a homogeneous liquid has been formed.

Paper: No. 1, sheets 25 cm square, with corner holes.

Locating reagents: (*a*) ninhydrin; (*b*) isatin; (*c*) Ehrlich reagent.

Tanks: cubic tanks 30 x 30 x 30 cm, with dural frame and solvent tray, each accommodating five paper sheets.

It is essential to handle the paper with the fingers as little as possible, especially when it is wet with solvents, as finger prints may show up on treatment with ninhydrin.

Take eight sheets of paper and draw two lines at right angles 2 cm from the edges. With the intersection (the origin) at the lower left-hand corner, mark the solvent flow directions – solvent 1 upward, and solvent 2 horizontally. On four further sheets of paper draw a line 2 cm from one edge, and mark seven points at equal intervals along it. These sheets are for one-way reference chromatograms of single amino-acids. Label all the sheets and origins.

Unscrew one side of the dural frame and remove all the spacers except one on each rod. Put one sheet on, and apply a spot of an amino-acid mixture solution at the origin – about 10 mm^3, giving about 25 μg of each acid; two drops from a platinum loop 4 mm in diameter will suffice (p. 145); the spot should not exceed 0.5 cm in diameter. Put on four spacers and then the next sheet; apply the same mixture. Repeat with the third and fourth sheets, but using a second mixture. The fifth should be one of the one-way sheets; apply one drop of each of seven individual amino-acid solutions (5 mm^3, giving about 12 μg) to the marked positions. Rinse and flame the platinum loop between applications. Finally replace the last spacers and the side panel and screw up carefully. Prepare two frames in this manner, using the same two mixtures, and the same seven acids on the one-way sheet. See that in the second frame the start line of the one-way sheet is correctly placed for running in solvent 2. Allow all the spots to dry.

Put 150 cm^3 of solvent in the tray in each tank (solvent 1 in one tank, and solvent 2 in the other). Put the frames in the trays with the appropriate starting lines lowermost, and replace the lids. It will be observed that no equilibration time is given, since reproducible results are possible without it. An overnight run of 12–16 hours should be allowed.

The following morning, remove the frames from the tanks, and dry the sheets by arranging the frame so that a current of air from a fan blows across the paper. Sheets which have been in solvent 1 will be dry in one hour, but those from solvent 2 must have at least four hours drying time.

When the sheets are dry, carefully unscrew the panel nearer to the one-way sheet, and remove that sheet. Replace it by a second one, ensuring that its starting line is located correctly at right angles to the starting line of the sheet just removed, and apply one spot of each of seven other amino-acids; replace the side panel. Repeat for the second frame, using the same seven acids.

Put fresh solvent in each tank, and allow each frame to have an overnight run in the solvent in which it was not run previously. Ensure that the frames are turned through 90°, so that the correct starting lines are lowermost. Remove and dry as before, dismantle the frames and hang the sheets up with clips.

It will be seen that there have now been obtained:
Two sheets containing one-way chromatograms of fourteen amino-acids run in solvent 1;
Two sheets containing one-way chromatograms of the same acids run in solvent 2;
Two sheets each containing a chromatogram of a mixture, run first in solvent 1, and then in solvent 2;

Two sheets each containing a chromatogram of the same mixture, run first in solvent 2, and then in solvent 1; and

Four sheets containing chromatograms of a second mixture, run under the same conditions.

The amino-acids are located by dipping. Hold the sheet by opposite edges with large 'bull-dog' clips, and draw it steadily with a see-saw motion through the reagent solution in a shallow dipping tray. Hang the paper up to dry by one of the clips. Treat the various sheets as follows:

One-way sheets: ninhydrin.

Two-way sheets for a particular mixture:

First sheet: (1) ninhydrin

             (2) Ehrlich reagent.

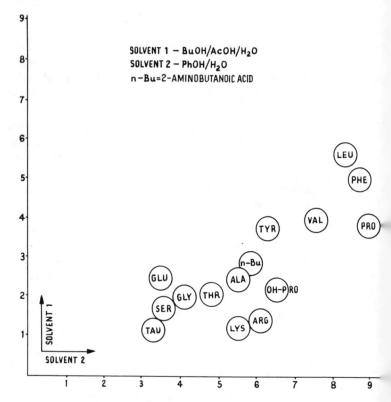

SOLVENT 1 − BuOH/AcOH/H_2O
SOLVENT 2 − PhOH/H_2O
n−Bu=2−AMINOBUTANOIC ACID

Fig. 6.5. 'Standard map' for separation of amino-acids (Experiment 13)

Second sheet: (1) isatin.

                (2) Ehrlich reagent.

After treatment with ninhydrin and with isatin, heat the sheet in the drying oven at 105°C for four minutes; mark all the spots which appear, examine the sheet under ultra-violet light, and identify as many of the acids as possible before applying the second reagent. Notes on the reactions of these reagents with amino-acids will be found in the Appendix.

Determine R_f values (Table 6.4) for the individual amino-acids from the one-way sheets, and from those, and the 'standard map' (Fig. 6.5) identify the constituents of the mixture. Notice that the one-way chromatograms only form a rough guide to the position of the substances on the two-way sheets, and also notice how the order in which the two solvents are used affects the two-way results.

It is not usually recommended partly to dismantle the frames between the two solvent runs in two-way chromatography; it is done here to save time, in that one-way as well as two-way chromatograms are produced.

Table 6.4 R_f *values for amino-acids (Experiment 13)*

	Solvent 1 BuOH/AcOH/H_2O	Solvent 2 PhOH/H_2O
alanine	0.24	0.55
2-aminobutanoic	0.28	0.58
arginine	0.13	0.60
glutamic acid	0.25	0.33
glycine	0.20	0.40
hydroxyproline	0.21	0.67
leucine	0.58	0.82
lysine	0.12	0.55
β-phenylalanine	0.50	0.86
proline	0.39	0.88
serine	0.19	0.34
taurine	0.12	0.33
threonine	0.21	0.49
tyrosine	0.38	0.62
valine	0.40	0.74

Suggested mixtures which give good two-way chromatograms are:

1	2	3
alanine	glutamic acid	2-aminobutanoic acid
arginine	glycine	leucine
glutamic acid	hydroxyproline	phenylalanine
proline	lysine	taurine
serine	proline	threonine
tyrosine	serine	tyrosine
valine		

Experiment 14: *Descending method*

Separation of sugars

Sugars can best be separated by descending chromatography. A long running time is used, because the R_f values are rather low, and those of similar sugars rather close, in all solvents. The solvent is allowed to run off the paper. This experiment, in addition to illustrating sugar chromatography, is intended to give experience in handling large sheets of paper; sugars can in fact be separated on smaller sheets in some cases. Synthetic mixtures of sugars are used here, for the same reasons as are given for amino-acids (Experiment 13). One-way runs are carried out in two different solvents, using two different locating reagents.

Solutions: reference solutions of individual mono-, di-, and oligo-saccharides (about 10% in water), and solutions of mixtures.

Solvents: (1) propan-1-ol/ethyl acetate/water 6 : 1 : 3 v/v.
(2) pyridine/ethyl acetate/water 2 : 2 : 1 v/v.

Paper: No. 1, sheets 46 x 57 cm.

Locating (*a*) silver nitrate/alcoholic alkali.
reagents: (*b*) aniline phthalate.

Tanks: large glass tanks 50 x 50 x 20 cm, each containing two solvent troughs. Four sheets can be put in each tank.

Put some of the appropriate solvent mixture in the two Petri dishes at the bottom of each tank about six hours before the run is started. Prepare two sheets of No. 1 paper 40 x50 cm, cut from the large sheets. The longer edge should correspond to the long edge of the original sheet (the machine direction, p. 127). The method of preparing the sheet is shown in Fig. 6.6. Draw a line 14 cm from one of the shorter sides, and parallel to it. Mark twelve points at equal intervals, with the end ones not less than 2 cm from the edge. Fold backward along CD and forward along EF. Cut the other end of the sheet to form a series of pointed tongues, to allow more even solvent flow from the bottom.

Take the long glass rods from the tank and put them in the drying rack (Fig. 3.22). Put the sheets on the rods along the fold CD, and secure them with wooden spring clips at the ends. Hold the sheet horizontal with one hand at the serrated end, and apply five reference solutions and one mixture in duplicate as indicated in Fig. 6.6 (M is the mixture). The sheet is later to be cut along AB. One drop from a platinum loop is sufficient for monosaccharides (40–50 μg), but two drops are needed for di- and oligo-saccharides.

When the spots are dry, put one sheet in each tank. To do that, lift the glass rod and sheet from the frame and put them in the tank with the fold EF in the empty trough and insert the heavy glass retaining rod along EF. Remove the clips and adjust the papers if necessary so that they hang straight. Replace the lid and leave for about thirty minutes. Then push the lid aside lengthways so that the ends of the troughs are just exposed, and carefully fill the troughs with the appropriate solvent mixture. Close the lid and allow an overnight run (12–16 hours).

At the end of the run, clip the sheets to the support rods again, cut

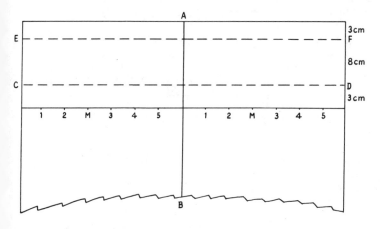

Fig. 6.6. Preparation of sheet for separation of sugars (Experiment 14)

away the paper dipping into the trough (see p. 147 and Fig. 3.21), and replace the rod and sheet in the drying rack. When the paper is dry (accelerated by means of an electric fan), cut the sheets along the line AB.

Dip one half of each sheet in silver nitrate solution, and then in alcoholic alkali. Reducing sugars give dark spots with the reagent, but the background darkens rather rapidly on exposure to air and light. The darkening can be restrained by cautiously dipping in 0.1 mol dm^{-3} sodium thiosulphate and then drying, but there may be some diffusion of the edges of the spots. Dipping is done in the large (50 cm) dipping tray. Spray the other half of each sheet with aniline phthalate reagent, and heat in the drying oven at $105°C$ for 5 minutes. The reagents and colours are described in the Appendix. Examine the sheets in ultra-violet light before and after they are treated with the reagents.

As the solvent runs off the sheet the R_f values cannot be determined; R_g values, based on glucose, are normally quoted, and thus glucose should always be one of the reference substances. The values should be measured, but the constituents of the mixtures are identified mainly by visual comparison of the chromatograms. Approximate R_g values for No. 1 paper are given in Table 6.5.

Suggested mixtures are:

1 glucose, lactose, sucrose, xylose
2 arabinose, maltose, raffinose
3 arabinose, galactose xylose

Table 6.5 R_g *values for sugars (Experiment 14)*

	Solvent 1 PrOH/EtOAc/H$_2$O	Solvent 2 Py/EtOAc/H$_2$O
arabinose	1.15	1.12
fructose	1.10	1.15
galactose	0.83	0.86
glucose	1.00	1.00
lactose	0.43	0.53
maltose	0.60	0.74
raffinose	0.38	0.40
sucrose	0.70	0.78
xylose	1.23	1.27

Experiment 15: *Horizontal method*

Rapid separation of amino-acids [16, 17, 18]

The Kawerau apparatus (p. 120) can be used to separate amino-acids where the R_f values are sufficiently different to make a two-way separation unnecessary, and large enough to allow a short run.

Solutions: cystine, hydroxyproline, β-phenylalanine, and mixtures of two or all (about 0.25% of each acid in water).

Solvent: butan-1-ol/acetic acid/water 12 : 3 : 5 v/v.

Paper: No. 4, type KCT, 14.5 cm (circles with five radial slits).

Locating reagent: isatin.

Tank: Kawerau apparatus.

Spots are applied at the inner apex of each sector. Apply five in all (three reference solutions and two mixtures). One drop from a platinum loop, giving about 12 μg, is adequate (p. 145). Allow the spots to dry. Put enough solvent in the dish to give a depth of about 0.5 cm, see that the capillary is running freely, and adjust its height to be just above the edge of the dish. Put the paper in place with its centre touching the capillary. If flow does not start, put a drop of solvent on the centre of the paper. When flow is established, replace the lid, and allow to run until the solvent front reaches the ends of the slits. Hold the sheet with forceps, dry it with warm air from the blower, and then dip it in the isatin solution in a large Petri dish. Drain, dry, and heat in the oven at 105°C for four minutes. The three amino-acids give characteristic colours with the reagent (see Appendix) and their order from the centre is cystine, hydroxyproline, β-phenylalanine (Fig. 6.7).

This separation can be carried out by the ascending method, using CRL/1 papers in the manner of Experiment 9.

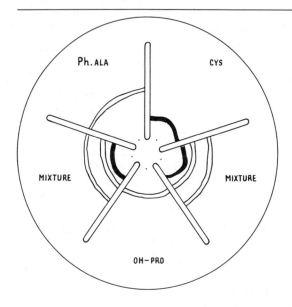

Fig. 6.7. Horizontal separation of amino acids (Experiment 15)

V Electrophoresis

Experiment 16: *Horizontal method – low-voltage paper electro-phoresis*

Separation of amino-acids

Electrophoresis is more often used to separate proteins than amino-acids. The degree of separation obtained for amphoteric electro-lytes depends on the pH of the buffer and on the iso-electric point of the substances. This experiment illustrates the use of a horizontal low-voltage paper method, which separates acidic, basic and neutral amino-acids but is not satisfactory for separating the members of one group from each other (p. 174).

Solution: a mixture of arginine, aspartic acid, and glycine (0.04 mol dm^{-3} with respect to each, in water).

Electrolyte
(buffer): 1 mol dm^{-3} acetic acid.
Paper: No. 1, reels 3 cm wide.
Locating
reagents: (*a*) ninhydrin; (*b*) isatin; (*c*) Sakaguchi reagent.
Apparatus: Horizontal electrophoresis tank. This experiment was originally designed for the Shandon Horizontal Electrophoresis Tank No. 2508, which is now superseded by the Model U77. Any horizontal electrophoresis apparatus could be used: in all cases manufacturers provide instructions for the use of their products.

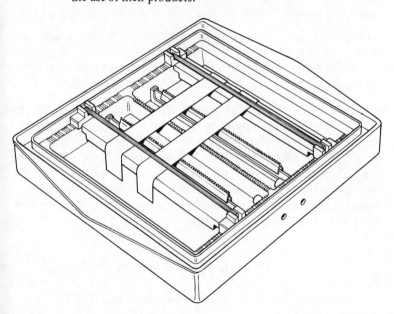

Fig. 6.8. Shandon horizontal electrophoresis apparatus U77 (Experiment 16)

Cut four strips of paper (they must all be exactly the same length) and mount them in the frame as described in the instructions. Number the strips, mark which end is to be in the anode compartment, and draw a line across each strip near that end, just clear of the paper support (put a thin glass strip under the paper while marking is being done).

Fill the tanks with buffer solution and put the frame in position. Moisten the strips by gently dabbing with pieces of paper dipped in the buffer solution, or with a small soft brush. Mop up any solution spilt on the tank or frame.

Adjust the tension of the strips by easing into position with the fingers at the ends, and then apply the test solution with a capillary tube along the lines to within 5 mm of the edges. Take up some of the solution in the capillary (melting point) tube and measure the exact length it occupies (7–10 mm). Apply the same volume to each strip.

Replace the lid, connect up the electrodes to the power pack, and apply a potential of 200 V for four hours; the current will be about 2 mA. Display the safety notice on the tank while the current is on, and do not touch the tank or connections without switching off.

At the end of the run, switch off, lift out the frame, and dry in a draught from a fan. Remove the strips from the frame, and dip two strips in ninhydrin solution and heat at 105°C for four minutes. All three acids are thus revealed, and the strips should appear identical. Dip the third strip in isatin solution and heat at 105°C for four minutes to detect aspartic acid, and treat the fourth strip with Sakaguchi reagent to detect arginine (see the Appendix for colour reactions). The order from the start is aspartic acid, glycine, arginine.

VI Gas Chromatography

Experiment 17: *A simple demonstration apparatus for GLC*

This apparatus (Fig. 6.9) was originally described by Cowan and Sugihara[19]. The column is made by bending a 1.5 m length of glass tubing of 6 mm bore, so that one limb is about twice the length of the other. To the shorter limb is attached a piece of wider tubing, packed with glass wool, having a serum cap at its upper end through which samples can be injected. A short piece of capillary tubing is attached to the column outlet, to serve as a jet for burning the effluent hydrogen.

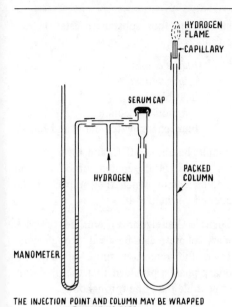

Fig. 6.9. Simple gas chromatography apparatus (Experiment 17)

A column packing which gives good results is the domestic detergent Tide. Thoroughly dry about 100 g of Tide, screen it, and collect the fraction between 60 and 100 mesh. Pack the column carefully, until, after prolonged tapping on the bench, no more material can be accommodated. Then join the column to the apparatus, turn on the hydrogen supply, and light the jet at the top of the column.

CAUTION. The flame is nearly invisible and it may be difficult to see if it is alight. Test from time to time with a scrap of paper.

A convenient sample size is $0.01-0.02$ cm^3, injected by means of an hypodermic syringe. The eluted substances are detected by the change in colour of the hydrogen flame as they emerge from the column. Best results can be obtained if the sample injection point is heated to about $100°$C by a small coil wrapped round the glass tubing. The column itself can be heated by wrapping with heating tape.

A typical separation is that of the four substances listed below, which are eluted in the order given:

	Flame colour
1. diethyl ether	bright yellow
2. carbon disulphide	light blue
3. benzene	luminous and smoky
4. tetrachloromethane	blue, changing to yellow and smoky

The column temperature should be about $50°C$, and the hydrogen flowrate should be $50–100 \text{ cm}^3 \text{ min}^{-1}$. Halogenated compounds can also be detected by the blue/green colour which they impart to the flame when a small coil of copper wire is held in it.

Other column packings, such as kieselguhr or crushed firebrick coated with different stationary phases, can be used. Rough comparisons can be made between the retention times of different substances on different stationary phases, provided that the operating conditions are kept as nearly as possible the same throughout.

Experiment 18: *GLC and GSC*

The experiments described here are best performed with an apparatus possessing a flowmeter and some form of detector, such as the simple katharometer described in the Appendix, coupled to a recorder. The column should be maintained at constant temperature. A simple column heater jacket can easily be constructed from a piece of glass tubing about 50 mm in diameter and 1 m long. First wrap the tube in a piece of asbestos paper anchored firmly in place, then wind the tube with nichrome wire. If the windings are closer together at the ends of the tube than in the middle a more uniform temperature will be maintained along the tube. Lag the tube with asbestos string, and finally, outside this, aluminium foil. Connect the winding to the mains through a variable transformer; with a little care it will be found possible to control the temperature within about $1°C$.

The arrangement in Fig. 6.10 will be found useful in that two different columns may be used at the same time in the same apparatus (see p. 197).

When using this arrangement it is convenient to set the base-line on

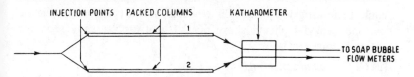

Fig. 6.10. Twin-column arrangement for use with katharometer

the recorder in the middle of the chart so that either column can be used without having to switch the signal from the katharometer.

(i) GLC – qualitative analysis

Pack U or W shaped columns of total length 1.5–2.0 m and bore 6 mm with Celite or firebrick (60–100 mesh) coated with (*a*) squalane (non-polar) and (*b*) dinonyl phthalate (moderately polar) stationary phases (see p. 222). Ether is a suitable solvent for coating purposes and the proportion of stationary phase on the solid when ready for packing should be about 20% w/w. If the apparatus in Fig. 6.10 is used column 1 could contain the dinonyl phthalate and column 2 the squalane. Experiments using the following substances can be run at room temperature without the disadvantage of very large retention volumes: diethyl ether, acetone, butanone, dichloromethane, trichloromethane, tetrachloromethane, 1,1-dichloroethane, 1,1,1-trichloroethane, *cis*-dichloroethylene, *trans*-dichloroethylene, 2-methylbutane, pentane, hexane, cyclohexane, and benzene.

Provided that there are no violent fluctuations in the room temperature, no special lagging of the columns is necessary. Students encountering gas chromatography for the first time seem to prefer to be able to see the columns but thermostating of the columns leads to more consistent results. A selection of, say, six of the above substances can be taken and about 1 mm^3 injected on each column in turn and the adjusted retention volumes, V_R, (p. 207), tabulated. An unknown mixture of three of the six substances can then be injected and the components identified by comparing the adjusted retention volumes with the tabulated values.

The advantage of performing the analysis on two different columns is that positive identification of the components is more certain and the effects of polarity in the stationary phases and the absorbates on the retention volume will also be illustrated. Other stationary phases which could be investigated include dimethyl sulpholane (very polar), silicone oil, and benzyl biphenyl above 60°C (slightly polar and useful for aromatic compounds).

(ii) GLC – quantitative analysis

The above experiment may be extended to yield quantitative results. The components of the mixture having been identified, their proportions may be determined fairly accurately by the following method which assumes that the detector response is a linear function of the concentration of the vapour – often reasonably so in the case of a katharometer. Inject fairly large samples of the pure components of the mixture, say 10 mm³, as accurately as possible, keeping the column conditions, especially flowrate of carrier-gas and temperature, as nearly as possible the same as for the separation. Carefully cut out the peaks from the recorder chart and weigh them. Compare the weights of these peaks with the corresponding peaks obtained from the mixture and hence determine the amounts of the various substances in the mixture. It will be seen that it is not necessary to know the amount of the mixture injected provided that comparison of the standard peaks is made with peaks obtained in the same separation.

(iii) GSC – qualitative analysis

Qualitative and quantitative analysis using columns packed with an active solid may be carried out in a similar fashion to that described above for GLC. However, if the same substances are to be injected the columns will need to be short (about 30 cm), and the experiments run at very much higher temperatures, 50–100°C. This is because of the much greater capacity of an active solid compared with an equal volume of GLC packing. In addition some substances such as ether on silica gel and acetone on alumina may remain indefinitely on the column or undergo chemical change. The real benefits of GSC are not evident unless separations are attempted on very volatile materials such as ethane, propane, butane, ethylene, carbon monoxide, and carbon dioxide. These substances can be separated on 1.5–2.0 m columns operated at room temperature. The same separations, even if they are possible, require much longer columns with GLC. Suitable column packings are silica gel, alumina, and active carbon, all activated at 200–300°C for 12 hours before use.

Pack two columns, 1.5–2.0 m long, one with silica gel and the other with active carbon, both freshly activated and 60–100 mesh, and assemble in an apparatus as shown in Fig. 6.10. When the columns are used at room temperature suitable samples for analysis are ordinary town gas which may contain – among other constituents – hydrogen, carbon monoxide, ethylene, ethane, propane, butane, and benzene; and Calor gas which consists mainly of butane, isobutane, and propane. The gas contained in the little cartridges sold for gas cigarette-lighters is similar to Calor gas. A convenient way of storing these gases immediately prior to an analysis is to use them to displace the air from a polythene wash-bottle and then close the wash-bottle with a rubber serum cap. Samples can then be drawn from the bottle by means of a hypodermic syringe of about 2 cm^3 capacity and injected directly onto the appropriate column. Unfortunately, pure samples of such gases as ethylene, ethane, and propane are rather expensive so that it may not be possible to identify all the constituents of a mixture by comparing retention times of peaks with those of pure substances. There are, however, mixtures of gases commercially available, analysed by gas chromatography which are convenient to use and less expensive than samples of the pure gases.

(iv) GSC – quantitative analysis

Quantitative analyses can also be carried out as described in Experiment 18 (*ii*) above, but when gases are being determined the accurate injection of a known volume of gas by means of a hypodermic syringe is not at all easy. The most satisfactory way of sample introduction for calibration purposes is to use a device such as that described on p. 187.

(v) GLC or GSC – measurement of HETP

The chromatograms obtained in Experiments 18 (*i*) and 18 (*iii*) may be used for the determination of the HETP for the given columns. The formulas given in Chapter 4 (p. 263) may be used. Under a given set of conditions on a given column the HETP should theoretically remain the same regardless of the nature of the substance whose peak is being used for the purpose of the calculation. In practice there will be some variation, particularly between substances of widely differing polarity. The HETP will depend, however, mainly on how well the column has been packed; values may range from a few millimetres to several centimetres.

If the peaks are markedly asymmetric – as may frequently occur in GSC – the HETP values are not very meaningful; only symmetrical or nearly symmetrical peaks should be used for the calculation.

(vi) GLC or GSC – determination of optimum flowrate of the carrier-gas

The van Deemter equation (p. 268) relates the HETP to the linear gas velocity ($\bar{u}$). The form of the equation is such that a curve with a minimum is obtained when HETP is plotted against $\bar{u}$. This minimum corresponds to the optimum gas velocity, and hence to the optimum flowrate because the value of $\bar{u}$ is obtained by dividing the column length by the time taken for the air peak to appear (p. 268).

From the results of experiments (*i*) and (*ii*) above choose a substance with a convenient retention time (about 10 minutes) under the column conditions prevailing and make a series of measurements of the HETP for a range of flowrates between 10 and 120 cm^3 min^{-1} in steps of 10 cm^3, that is, 10, 20, 30, etc. Plot the flowrate (or linear gas velocity) against the HETP.

(vii) GLC – separation of homologous series

Take a homologous series of lower alcohols, ethers, ketones, or esters and determine the retention time of each member. Pot log V_R, against the carbon number (p. 211).

VII Thin-layer Chromatography

The experiments described in this section can be carried out on plates prepared by any convenient method, and therefore no particulars of apparatus are given. It is assumed, however, for convenience in giving experimental instructions, that standard 20 x 20 cm plates are used, and details of plate preparation, other than the method of spreading, are given in each case. In one-way separations the prepared plates should be used as follows:

(1) To obtain a sharply defined edge, remove about 5 mm of adsorbent from each side of the plate with a finger or spatula.
(2) Mark the position of the ends of the start line by making small marks with a sharp hard pencil.

(3) To ensure that the solvent flow stops when the front has travelled a pre-determined distance, draw a line across the plate with a sharp hard pencil. Gently remove the small amount of displaced adsorbent by blowing. The plate will then appear as in Fig. 5.8 (p. 296).

(4) Apply the unknown and the reference solutions, with a pipette or other means, at regular intervals along the start lines. Spots should be not less than 1 cm apart and not nearer than 1 cm to the edge. It is essential not to disturb the layer when applying samples. Allow the spots to dry.

For operations (2), (3), and (4) above a special template may be used as a guide, or a ruler may be supported on either side of the plate so that it is just clear of the adsorbent layer.

(5) Place the plates in an appropriate tank, which contains mobile phase to a depth of 0.5–1.0 cm. If small tanks are used, no equilibration is needed for the experiments described, but the lid must fit properly.

(6) When the mobile phase reaches the finishing line, remove the plates from the tank, allow them to dry, and then apply the final treatment, if necessary.

Experiment 19: *Adsorption chromatography – ascending method*

Separation of aldehyde 2,4-dinitrophenylhydrazones

This experiment may be compared with experiment 12, where separation of the same substances by reversed phase paper chromatography is described.

Solutions: see page 334.
Solvent: benzene/light petroleum (60–80) 3 : 1 v/v.
Plates: silica gel with binder (such as silica gel G), 250 μm thick, made up with water and dried at 100°C for 30 minutes.

Mark out the plate as described above, and apply 2 mm^3 of each reference solution and of the mixture at even intervals along the start line. Allow a 15 cm run.

When the run is finished, dry the plates, and observe the position of the spots in daylight and in ultra-violet light. Approximate R_f values are given in Table 6.6.

Table 6.6   *Approximate R_f values for Experiment 19*

2,4-dinitrophenylhydrazone	R_f
formaldehyde	0.30
acetaldehyde	0.26
propionaldehyde	0.39
butyraldehyde	0.45

Experiment 20: *Adsorption chromatography – ascending method*

Separation of a mixture of dyes [20]

This experiment is designed to show the effect of changes in the polarity and pH of the mobile phase on R_f values.

Solution: a convenient solution is the test mixture described on p. 304 (which is supplied with one form of commercial apparatus). It consists of 1% w/v of indophenol blue, Sudan red G, and *p*-dimethylaminoazobenzene in benzene.

Solvents: 1. chloroform/ethyl acetate/formic acid 5 : 4 : 1 v/v.
2. toluene/ethyl formate/formic acid 5 : 4 : 1 v/v.
3. benzene/chloroform 1 : 1 v/v.
4. benzene/ethyl formate/formic acid 75 : 24 : 1 v/v.

Plates: silica gel with binder, 250 μm thick, made up with water and dried at 100°C for 30 minutes.

(*a*) Apply ten 4 mm^3 spots of the solution on each of four plates, and run one plate in each solvent, allowing a 10 cm run in each case.

(*b*) Select the solvent which appears to give the best separation and apply a series of spots of different amounts (2, 4, 6, etc. mm^3). When more than 4 mm^3 is being applied the operation should be done in portions of not more than 4 mm^3 at a time, the spot being allowed to dry between applications. Allow a 10 cm run as in (*a*).

After the runs, allow the plates to dry, and compare the R_f values of the dyes in the various solvents.

In (*a*) observe whether there are any significant edge effects, or irregularities due to variations in the layers, and in (*b*) observe whether there is a reasonable quantitative difference between the area of the spots, or their depth of colour. Approximate R_f values are given in Table 6.7.

Table 6.7 *Approximate R_f values for Experiment 20*

Substance	Colour of spot	Solvent			
		1	2	3	4
p-dimethylaminoazobenzene	yellow	0.15	0.28	0.65	0.71
Sudan red G	red	0.70	0.68	0.00	0.65
indophenol blue	blue/pink	0.06	0.05	0.00	0.28

Note that the three substances do not always appear in the same order, and that the colours vary somewhat according to pH.

Experiment 21: *Partition chromatography – ascending method*[21]

Separation of simple sugars

This experiment illustrates an application of partition chromatography on thin layers. The plates are made up from a slurry with aqueous sodium acetate (instead of water), which causes the retention of a relatively large amount of water when the plates are 'dried'. This deactivates the silica as an adsorbent, but is sufficient to form a polar stationary liquid phase.

Solutions: simple sugars, such as fructose, glucose, mannose, sucrose, and xylose, in pyridine (1% w/v), and mixtures of two or more.

Solvent: ethyl acetate/propan-2-ol/water 26 : 14 : 7 v/v.

Plates: silica gel with binder, $250\,\mu\text{m}$ thick, made up with $0.2\,\text{mol dm}^{-3}$ sodium acetate instead of water, and dried at $100°\text{C}$ for 30 minutes.

Locating reagent: anisaldehyde.

Apply 2 mm^3 spots of each solution, and allow a run of 15 cm. At the end of the run, allow the plates to dry, and spray with anisaldehyde reagent. Heat at 100°C for 10 minutes. Measure R_f values and identify the constituents of the mixture. Approximate R_f values are given in Table 6.8, and the colour reactions of sugars with the reagent are given in the Appendix.

Table 6.8 *Approximate R_f values for sugars (Experiment 21)*

	R_f
fructose	0.34
glucose	0.37
mannose	0.40
sucrose	0.23
xylose	0.53

Note that the spots are more diffuse than is usual in thin-layer chromatography, because a partition system is being used, and the results are therefore rather similar to those obtained on paper.

Experiment 22: *Adsorption chromatography — two-way method* [22]

Separation of amino-acids

Simple mixtures of amino-acids may be separated very quickly by two-way thin-layer chromatography. The procedure is similar to that used in paper chromatography (see Experiment 13).

Solution: mixture 1 on p. 339 (0.25% of each acid in water or 10% propan-2-ol).

Plates: silica gel with binder, 250 μm thick, made up with water and dried at 100°C for 30 minutes.

Solvent: 1. chloroform/methanol/17% aqueous ammonia 2 : 2 : 1 v/v.

        2. phenol/water – as solvent 2 in Experiment 13 (see
        p. 336).

Locating
  reagent: ninhydrin.

Mark two lines at right angles on the plate, 3 cm from the edges, as
shown in Fig. 5.9 (p. 297), so that a run of 15 cm in each direction is
obtained if the start is at the point indicated. Note that the two edges
of the plate marked A and B must be free from irregularities and
blemishes, and the plate must be handled only by the opposite edges.
Apply a total of 10 mm³ of the solution to the start point. As

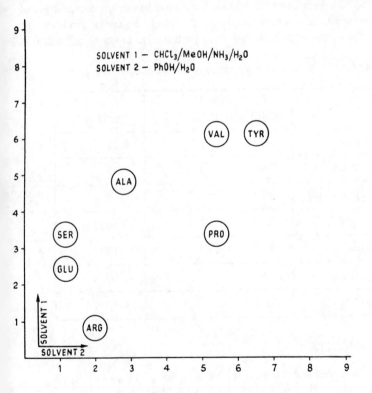

*Fig. 6.11. 'Standard map' for separation of amino-acids
(Experiment 22)*

described in Chapter 5, p. 294, not more than 5 mm^3 must be applied at a time. The solvent should be 1 cm deep in each tank.

Allow to run in solvent 1 until the solvent front reaches the finishing line. Remove the plate from the tank, allow the first solvent to evaporate in the atmosphere, and then put the plate (turned through 90°) in the tank containing the second solvent. At the end of the second run, remove the plate from the tank and allow the solvent to evaporate from it. A very gentle current of warm air passing over the plate laid horizontally may be used to accelerate the removal of the phenol, which is essential if the locating reagent is to work properly.

When the plate is free of phenol, spray it with ninhydrin, and heat it at 105°C for about 4 minutes (see the Appendix for the composition of the reagent, and colour reactions). A 'standard map' is given in Fig. 6.11, and approximate R_f values in each solvent are given in Table 6.9.

Table 6.9   *Approximate R_f values of amino-acids (Experiment 22)*

	Solvent 1 CHCl$_3$/MeOH/aq.NH$_3$	Solvent 2 PhOH/H$_2$O
alanine	0.47	0.25
arginine	0.06	0.18
glutamic acid	0.22	0.10
proline	0.32	0.50
serine	0.30	0.10
tyrosine	0.60	0.55
valine	0.60	0.44

REFERENCES

1. Cook. A. H., *J. Chem. Soc.*, 1938, 876.
2. Cook, A. H., Jones, D. G., *ibid.*, 1938, 1309.
3. Wilson, J. M., Newcombe, R. J., Denaro, A. R., Rickett, R. M. W., *Experiments in Physical Chemistry*, Pergamon Press, London, 2nd Edn., 1968.

4. Burstall, F. H., Davies, G. R., Wells, R. A., *Discuss. Faraday Soc.*, 1949, **7**, 179.
5. Salmon, J. E., Hale, D. K., *Ion Exchange, A Laboratory Manual*, Butterworth, London, 1959, p. 120.
6. Riemann, W., Lindbaum, S., *Analyt. Chem.*, 1952, **24**, 1199.
7. Burstall, F. H., Davies, G. R., Linstead, R. P., Wells, R. A., *J. Chem. Soc.*, 1950, 516.
8. Hunt, E. C., North, A. A., Wells, R. A., *Analyst*, 1955, **80**, 172.
9. Jakubovic, A. O., Knight, C. S., *Chromatographic and Electrophoretic Techniques*, Ed. Smith, I., Heinemann, London, 1960, **1**, p. 559.
10. Seligman, R. B., Edwards, M. D., *Chem. and Ind.*, 1955, 1406.
11. Consden, R., Gordon, A. H., Martin, A. J. P., *Biochem. J.*, 1944, **38**, 224.
12. Dent, C. E., *ibid.*, 1948, **43**, 169.
13. Jepson, J. B., Smith, I., *Nature*, 1953, **171**, 43 and **172**, 1100.
14. Partridge, S. M., *Biochem. J.*, 1948, **42**, 238.
15. *idem, Nature*, 1949, **164**, 143.
16. Giri, K. V., Rao, N. A. N., *ibid.*, 1952, **169**, 923.
17. Proom, H., Woiwod, A. J., *J. Gen. Microbiol.*, 1951, **5**, 681.
18. Kawerau, E., *Biochem. J.*, 1956, **64**, 32P.
19. Cowan, P. J., Sugihara, J. M., *J. Chem. Educ.*, 1959, **36**, 246.
20. Stahl, E., Schorn, P. J., *Z. physiol. Chem.*, 1961, **325**, 263.
21. Stahl, E., Kaltenbach, U., *J. Chromatog.*, 1961, **5**, 351.
22. Fahmy, A. R., Niederweiser, A., Pataki, G., Brenner, M., *Helv. Chim. Acta*, 1961, **44**, 2022.

Appendix

1. Location reagents for paper and thin-layer chromatography

The methods of making up and using the location reagents mentioned in the experiments described in Chapter 6 are given in this Appendix. Only the reactions of those substances actually mentioned are listed here. Most of the reagents are, in fact, of far wider application. The works listed in the Bibliography (p. 372) should be consulted for fuller details.

Experiments 7, 9, and 11

(a) *rubeanic acid/salicylaldoxime/alizarin (RSA):* 0.1% w/v of each in ethanol. Spray the paper, and expose to ammonia while still damp. Allow the purple background colour to fade.

(b) *diphenylcarbazide:* 1% w/v in ethanol. Spray, expose to ammonia while damp, and allow the background colour to fade.

(c) *sodium pentacyanoammineferrate*(II)/*rubeanic acid (PCFR):* mix 0.7 g of the sodium salt in 20 cm³ water with 0.25 g rubeanic acid in 10 cm³ ethanol; shake for 15 minutes and filter. (Keeps for only 24 hours). Expose the paper to ammonia and spray with the reagent. Wash the paper in a shallow dish with 2 mol dm⁻³ acetic acid until all the background colour has cleared, and then with water.

(d) *rubeanic acid:* 0.1% w/v in ethanol/water 60 : 40 v/v. Expose the paper to ammonia and then spray with the reagent.

(*e*) *ammonium sulphide:* dip the paper in yellow ammonium sulphide solution, or dip in hydrogen sulphide solution and then expose to ammonia, or expose the damp sheet to ammonia vapour and then to gaseous hydrogen sulphide.

The colour reactions of the metals mentioned in the experiments with the appropriate reagents are listed in Table A.1.

Table A.1 *Colour reactions of metals*

Reagent	(*a*) RSA	(*b*) diphenyl-carbazide	(*c*) PCFR	(*d*) rubeanic acid	(*e*) ammonium sulphide
bismuth		pink	brown	yellow/brown	black
cadmium			blue	yellow	yellow
cobalt	orange	purple	green/brown	orange	black
copper	green		brown/black	green	black
iron	purple/brown	pink	blue/green	brown	black
lead		purple	purple/grey		black
manganese	pale brown	pale pink			
mercury		purple/pink	grey/green		black
nickel	blue	red	blue	blue	black
zinc	purple	pink	red		

Experiment 8

 (*f*) *silver nitrate/fluorescein:* 0.05 mol dm^{-3} aqueous silver nitrate/saturated solution of fluorescein in ethanol 96 : 4 v/v. Spray or dip the paper, wash by dipping in 2 mol dm^{-3} nitric acid, and then in water. Dark bands on a fluorescent background under ultra-violet light indicate chloride, bromide, and iodide. The chloride band darkens rapidly, and the iodide band may appear pale yellow in white light.

 (*g*) *zirconium/alizarin S:* equal volumes of 0.1% zirconium nitrate in concentrated hydrochloric acid/water 10 : 50 v/v, and 0.1% sodium alizarin S in water. Spray or dip the paper. A pale yellow spot on a pink background indicates fluoride.

Experiments 13, 15, and 16

 (*h*) *ninhydrin:* 0.2% w/v in acetone; add a few drops of pyridine before using (by dipping). This reagent gives colours with a large number of substances besides amino-acids. The colours appear after some hours at room temperature, but more quickly on heating (105°C for 4 minutes). The longer period at lower temperature reduces the colours produced by other substances.

 (*i*) *isatin;* 0.2% in acetone (use by dipping). A reagent of rather less versatility than ninhydrin, but useful in some specific cases. Requires heating to bring up the colours (105°C for 4 minutes).

 (*j*) *Ehrlich reagent:* 10% *p*-dimethylaminobenzaldehyde in hydrochloric acid. 1 volume of this stock solution is diluted with 4 volumes of acetone immediately before use (by dipping). Being strongly acid, this reagent destroys the colours produced by ninhydrin and isatin, but in a few instances characteristic changes occur.

Table A.2   *Colour reactions of amino-acids*

Reagent	(h) ninhydrin	(i) isatin	(k) Sakaguchi	Ehrlich (j) after ninhydrin (h)	Ehrlich (j) after isatin (i)
alanine	purple	pale grey		pale yellow	
2-amino-butanoic acid	purple	pale blue			
arginine	purple	v. pale pink	orange	yellow	
aspartic acid	purple/brown	blue			
cystine	purple	purple			
glutamic acid	purple	pale brown			
glycine	purple	pale pink			
hydroxyproline	yellow	blue		pale pink	purple/red
leucine	purple				v. pale pink
lysine	purple	purple		yellow/brown	
β-phenyl-alanine	purple	blue/grey			
proline	yellow	grey		pink	
serine	purple				
taurine	purple			pale yellow	
threonine	purple				
tyrosine	purple			v. pale orange	
valine	purple				

(k) *Sakaguchi reagent:* (1) 0.1% w/v oxine in acetone
(2) 0.3 cm³ bromine in 100 cm³ 0.5 mol dm⁻³ aqueous sodium hydroxide.

Dip the paper in solution (1), and allow the acetone to evaporate; then dip in solution (2) and allow to dry. This reagent is a general one for substituted guanidines, and is used here specifically for arginine.

The reactions of the amino-acids mentioned in Experiments 13, 15, and 16 with reagents (h), (i), (j), and (k) are listed in Table A.2.

Table A.3 *Colour reactions of sugars*

Reagent	(l) silver nitrate	(m) aniline phthalate	(n) anisaldehyde
arabinose	grey	red	
fructose	grey	brown	
galactose	grey	brown	
glucose	grey/black	brown	blue
lactose	brown	brown	
maltose	brown	brown	
mannose	grey	brown	green
raffinose		brown	
sucrose		brown	blue
xylose	grey	red	green/grey

Experiment 14

(*l*) (1) *silver nitrate:* saturated solution in water, 1 volume/acetone, 100 volumes.

(2) *sodium or potassium hydroxide:* 0.5% w/v in ethanol.

(3) *sodium thiosulphate:* 0.1 mol dm^{-3} in water.

Dip the paper in solutions (1), (2) and (3) in succession, allowing each solvent to evaporate before dipping in the next solution. Dip in (3) as quickly as possible, but with care; dipping in aqueous solutions is not recommended usually because the paper becomes so tender.

(*m*) *aniline phthalate:* 4.65 g aniline and 8.0 g phthalic acid in 500 cm^3 butan-1-ol saturated with water. Spray the paper, and heat it at 105°C for five minutes.

The reactions of the sugars mentioned in Experiment 14 with reagents (*l*) and (*m*) are listed in Table A.3.

Experiment 21

(*n*) *anisaldehyde:* 5 cm^3 anisaldehyde and 5 cm^3 conc. sulphuric acid in 90 cm^3 ethanol/water 95 : 5 v/v. Spray the plate and heat at 100°C for ten minutes. The colour reactions of the sugars mentioned in Experiment 21 are listed in Table A.3.

Experiment 22

(*o*) *ninhydrin:* 0.2% w/v in chloroform/butan-1-ol 50 : 50 v/v. Spray the plate and heat at 105°C for four minutes. The colour reactions of amino-acids with reagent (*o*) are the same as those with reagent (*h*), given in Table A.2.

2. A simple demonstration experiment in
paper chromatography

A very simple demonstration of paper chromatography can be made by chromatographing the ink of ball-point pens. Any of the small scale methods can be used (gas jar, Petri dish, etc.), and the initial spot is made by marking a spot or line with the pen itself. A suitable solvent is methanol/water/conc. HCl 82 : 9 : 9 v/v. A separation can be seen in about five minutes with many inks, and of course no location reagent is needed.

3. Apparatus for gas chromatography

(a) A very simple katharometer

The simple cell illustrated in Figs. A.1 and 4.10(b) (p. 196) is quite easy to construct, and, if care is taken in its use, is capable of reasonable performance. The dimensions are only approximate. The guiding principles are to keep the gas

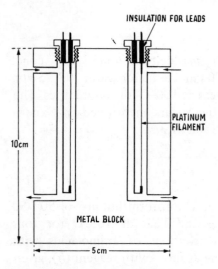

INSULATION FOR LEADS

PLATINUM FILAMENT

10cm

METAL BLOCK

5cm

Fig. A.1. Simple katharometer

(*a*) top view

(*b*) bottom view

Fig. A.2 Flame ionisation detector
By courtesy of Associated Octel Co. Ltd

volumes of the cells as small as possible, and to make the two cells as nearly identical as possible. The sensing elements should be constructed first. The brass plugs can be threaded, or sealed in place with wax or Araldite. Each has two holes drilled in it to take two stout wires as shown. To insulate the wires from the plug, each is wrapped with a small piece of plastic adhesive tape where it fits into the plug. Final fixing of the wires can be done with Araldite cement. The element itself can be made from platinum wire 0.05 mm in diameter, and about 30 cm long (cold resistance about 20 Ω). Coil the wire by winding it round a thick piece of steel wire, and solder it into position so that the coil is under slight tension.

The brass block may be flat or cylindrical. When the sensing elements have been made, drill holes in the block of the appropriate size and fix the elements in. Two elements are necessary in order to make the cell less sensitive to fluctuations in flowrate and ambient temperature. The column effluent passes through one cell, and carrier gas from a dummy column passes through the other. A suitable bridge current is about 150 mA. For the best results the cell should be maintained at a constant temperature; for simple investigations no thermostat arrangement is necessary, but the brass block should be lagged with insulating material.

(b) A flame ionisation detector*

The detector is illustrated in Fig. A.2. The combustion chamber about 4 cm in diameter, is made from brass and it fits directly on to the column. The burner jet is a hypodermic needle (Record No. VI), mounted on a short length of glass tubing which has been ground to the appropriate taper. The tubing is cemented into the base of the combustion chamber where it fits on to the column, and it also serves to insulate

---

* The details of this apparatus were supplied by Mr. G. F. Harrison, of The Associated Octel Co., Ltd., and are published with his permission.

the jet from the chamber. Two platinum electrodes 5 mm square are mounted 7 mm apart one on each side of the burner. The leads to the electrodes pass through the base of the combustion chamber and are insulated from it by the electrode mountings. Air, from a compressed air cylinder, enters the chamber through the copper pipe shown in the figure at the rate of about 1 dm³ min⁻¹. In the base of the chamber there is a 1 cm layer of glass beads to prevent air turbulence, which would produce noise in the detector. A battery-operated simple impedance matching circuit is shown in Fig. A.3. It should be used in conjunction with a 10 mV recorder.

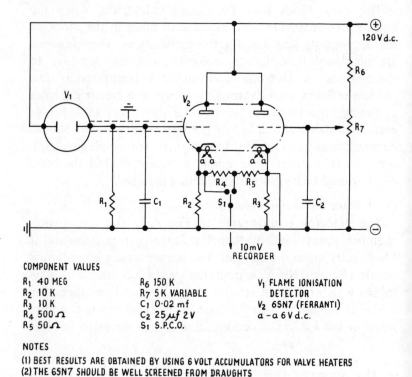

COMPONENT VALUES

R₁ 40 MEG	R₆ 150 K	V₁ FLAME IONISATION
R₂ 10 K	R₇ 5 K VARIABLE	DETECTOR
R₃ 10 K	C₁ 0·02 mf	V₂ 6SN7 (FERRANTI)
R₄ 500 Ω	C₂ 25 μf 2 V	a – a 6 V d.c.
R₅ 50 Ω	S₁ S.P.C.O.	

NOTES

(1) BEST RESULTS ARE OBTAINED BY USING 6 VOLT ACCUMULATORS FOR VALVE HEATERS
(2) THE 6SN7 SHOULD BE WELL SCREENED FROM DRAUGHTS

Fig. A.3. Flame ionisation detector—circuit diagram

4. Visual aids

The following is a selection of commercially-available visual aid material dealing with general aspects of chromatography. The list is not exhaustive, and there is other material which deals with specific separations. There are also films and strips on ion-exchange resins, but they are largely confined to non-chromatographic uses.

Audio courses

Gel permeation chromatography – J. Cazes. Four cassettes (3.7 hours) and course manual. American Chemical Society study unit series. Education Department, ACS, 1155 16th Street, NW, Washington D.C. 20036, USA.

Tape-slide

Series 2000 Basic gas chromatography – H. M. McNair. Six programme set.
Advanced gas chromatography – H. M. McNair. Two programmes.
Series 4000 Basic liquid chromatography – K. J. Bombaugh. Five programme set. Pattison & Co., St. Albans, Herts.

Films

Chromatography – Colour. Sound. 16 mm. 1964. Imperial Chemical Industries Ltd. Loan from ICI Film Library, Imperial Chemical House, Millbank, London.
Gas chromatography – Colour. Sound. 16 mm. 25 min. 1962. Perkin-Elmer Corp., USA.

Film loops

The mechanism of chromatographic separation – Standard 8 mm. Longman Group Ltd., Burnt Mill, Harlow, Essex.

Application of paper chromatography – Standard 8 mm. Nuffield Advanced Science Series, Penguin Education, Harmondsworth, Middx.

Paper chromatography – Super 8 mm. Thorne Films, 1229 University Avenue, Boulder, Colorado, 80302, USA.

Paper chromatography – Super 8 mm. Address as below.

Thin-layer chromatography – Super 8 mm.

Both from Gateway Educational Films Ltd., 29/31 Broad Street, Bristol BS1 2HF.

Bibliography

The following is a list of specialist journals and a selection of recently published texts

Journals

Gas and Liquid Chromatography Abstracts*; published quarterly; Applied Science Publishers, Barking, Essex, England (see Reference 1 p. 272).

Journal of Chromatography, M. Lederer (Ed.), published fortnightly; Vol. 1, 1958; Elsevier, Amsterdam.

Chromatographic Reviews, M. Lederer (Ed.), published annually; Vol. 1, 1959; Elsevier, Amsterdam.

Journal of Chromatographic Science (formerly Journal of Gas Chromatography), published monthly; Vol. 1, 1963; Preston Technical Abstracts Co., Evanston, Illinois.

Chromatographia; published monthly; Vol. 1, 1968; Friedr. Vieweg & Sohn GmbH, Braunschweig, W. Germany.

General, adsorption and partition chromatography

Chromatography, E. Heftman (Ed.), Reinhold, New York, 2nd Edition, 1967.

Practical Chromatographic Techniques, A. H. Gordon and J. E. Eastoe, Newnes, London, 1964.

An Introduction to Chromatography, D. Abbott and R. S. Andrews, Longmans, London, 1965.

---

* Circulated free to members of the Chromatography Discussion Group. For details of membership write to: The Executive Secretary, Chromatography Discussion Group, Trent Polytechnic, Burton Street, Nottingham NG 1 4BU, England.

Introduction to Chromatography, J. M. Bobbitt, A. E. Schwarting and R. J. Gritter, Van Nostrand Reinhold, London and New York, 1968.

Laboratory Handbook of Chromatographic Methods, O. Mikes, Van Nostrand, London and New York, 1966.

Chromatography, D. R. Browning (Ed.), McGraw-Hill, London, 1969.

Technical Dictionary of Chromatography, H-P. Angele, Pergamon, Oxford, 1970.

Advances in Chromatography, J. C. Giddings and R. A. Keller (Eds.), Vols. 1–10, Marcel Dekker, London and New York; Vol. 1, 1965; Vol. 10, 1970.

Principles of Adsorption Chromatography, L. R. Snyder, Edward Arnold, London, Marcel Dekker, New York, 1968.

Comprehensive Analytical Chemistry, C. L. Wilson and D. W. Wilson (Eds.), Vol. IIB Physical Separation Methods: Liquid Chromatography in Columns, Gas Chromatography, Ion Exchangers, Distillation; Elsevier, Amsterdam, London and New York, 1968.

Practical Liquid Chromatography, S. G. Perry, R. Amos and P. I. Brewer, Plenum Press, New York, 1972.

Modern Practice of Liquid Chromatography, J. J. Kirkland (Ed.), Wiley-Interscience, New York and London, 1971.

Paper and thin-layer chromatography

Chromatographic and Electrophoretic Techniques; Vol. I, Chromatography, I. Smith, Heinemann, London, 3rd Edition, 1968.

Paper Chromatography, A Comprehensive Treatise, I. M. Hais and K. Macek, Academic Press, New York and London, 1963.

Bibliography of Paper Chromatography and Survey of Applications, 1960–69, (Journal of Chromatography Supplementary Volume No. 2), K. Macek *et al,* Elsevier, Amsterdam, London and New York, 1972.

Quantitative Paper and Thin-Layer Chromatography, E. J. Shellard (Ed.), Academic Press, New York and London, 1968.

Paper and Thin-Layer Chromatography, I. Smith and J. G. Feinberg, Longmans, London, 2nd Edition, 1972.

Thin-Layer Chromatography, K. Randerath, Academic Press, New York and London, 2nd Edition, 1966.

Thin-Layer Chromatography, E. Stahl (Ed.), Allen and Unwin, London, 2nd Edition, 1969.

Thin-layer Chromatography, J. G. Kirchner, Wiley, New York and London, 1967.

Thin-Layer Chromatography; an Annotated Bibliography, 1964–68, B. J. Haywood, Ann Arbor-Humphrey, London, 1968.

An Introduction to Chromatography on Impregnated Glass Fibre, F. C. Haer, Ann Arbor-Humphrey, London, 1969.

Gel chromatography

Gel Chromatography, H. Determan, Springer Verlag, New York, 1968.

An Introduction to Gel Chromatography, L. Fischer, North-Holland, London, 1969.

Electrophoresis

Chromatographic and Electrophoretic Techniques, Vol. 2: Zone Electrophoresis, I. Smith, Heinemann, London, 3rd Edition, 1968.

Electrophoresis in Stabilising Media, G. Zweig and J. R. Whitaker, Academic Press, London and New York, 1967.

Methods of Zone Electrophoresis, J. R. Sargent, British Drug Houses, Poole, 2nd Edition, 1969.

Electrophoresis–Technical Applications, B. J. Haywood, Ann Arbor-Humphrey, London, 1969.

Ion-exchange

Synthetic Ion Exchangers, G. H. Osborn, Chapman & Hall, London, 2nd Edition 1961.

Ion Exchange, F. G. Helfferich, McGraw-Hill, New York, 1962.

Inorganic Ion Exchange Materials, C. B. Amphlett, Elsevier, Amsterdam, 1964.

An Introduction to Ion-Exchange, R. Paterson, Heyden, London, 1969.

Gas chromatography

Vapour Phase Chromatography, D. H. Desty (Ed.), Butterworths, London, 1956.

Gas Chromatography 1958, D. H. Desty (Ed.), Butterworths, London, 1958.

Gas Chromatography 1960, R. P. W. Scott (Ed.), Butterworths, London, 1960.

Gas Chromatography 1962, M. van Swaay (Ed.), Butterworths, London, 1962.

Gas Chromatography 1964, A. Goldup (Ed.), Institute of Petroleum, London, 1965.

Gas Chromatography 1966, A. B. Littlewood (Ed.), Institute of Petroleum, London, 1967.

Gas Chromatography 1968, C. L. A. Harbourn (Ed.), Institute of Petroleum, London, 1969.

Gas Chromatography 1970, R. Stock (Ed.), Institute of Petroleum, London, 1971.

Gas Chromatography 1972, S. G. Perry (Ed.), Applied Science Publishers, Barking, England, 1973.

Open Tubular Columns in Gas Chromatography, L. S. Ettre, Plenum Press, New York, 1965.

Ancillary Techniques of Gas Chromatography, L. S. Ettre and W. H. McFadden, (Eds.), Wiley-Interscience, New York and London, 1969.

A Programmed Introduction to Gas-Liquid Chromatography, J. B. Pattison, Heyden & Son, London, 1969.

Gas Chromatography, Principles, Techniques and Applications, A. B. Littlewood, Academic Press, New York, 2nd Edition, 1970.

Gas Chromatography, D. Ambrose, Butterworths, London, 2nd Edition, 1971.

Gas Liquid Chromatography, D. W. Grant, Van Nostrand-Reinhold, New York and London, 1971.

Recent Advances in Gas Chromatography, I. I. Domsky and J. A. Perry (Eds.), Marcel Dekker, New York, 1971.

Gas Analysis by Gas Chromatography, P. G. Jeffrey and P. J. Kipping, Pergamon Press. Oxford, 2nd Edition, 1972.

Chromatographic Systems—Maintenance and Trouble Shooting, J. Q.

Walker, M. T. Jackson and J. B. Maynard, Academic Press, London and New York, 1972.

Gas-Chromatographic Analysis of Trace Impurities, V. G. Berezkin and V. G. Tatarinskii, Consultants Bureau, New York and London, 1973.

Identification of Organic Compounds with the aid of Gas Chromatography, R. C. Crippen, McGraw-Hill, New York and London, 1973.

Index